BIOCHEMISTRY RESEARCH TRENDS

BIOCHEMISTRY AND MOLECULAR BIOLOGY

THE COMPLEXITY OF HUMAN TRAITS AND DISEASES

BIOCHEMISTRY RESEARCH TRENDS

Additional books in this series can be found on Nova's website under the Series tab.

Additional e-books in this series can be found on Nova's website under the e-book tab.

BIOCHEMISTRY RESEARCH TRENDS

BIOCHEMISTRY AND MOLECULAR BIOLOGY

THE COMPLEXITY OF HUMAN TRAITS AND DISEASES

MARY MIU YEE WAYE, PHD
EDITOR

nova publishers
New York

Library of Congress Cataloging-in-Publication Data

ISBN: 978-1-63482-312-8

Published by Nova Science Publishers, Inc. † New York

CONTENTS

INTRODUCTION

*Mary Miu Yee Waye, PhD**
School of Biomedical Sciences
President, Hong Kong Society of Biochemistry and Molecular Biology
Founding Director, Croucher Laboratory for Human Genomics
The Chinese University of Hong Kong
Hong Kong

Even though the ApoA5 gene was discovered more than 13 years ago by Pennacchio et al, 2001 [1], and it was known then the gene influences triglycerides, leading to speculations that it would affect the risk of stroke; the factors that affect cardiovascular diseases are very complex. Larry Baum's group shown that APOA5 genotypes were not significantly associated with strokes or stroke subtypes in the cohort they studied [2]. The discovery of the ApoA5 gene was very special in that, unlike many other genes discovered through studying a known protein or enzyme, advanced bioinformatics was used for the initial gene discovery, an amazing feat that could not have been done without knowledge of the human genome. The laborious effort in deciphering the genome was very controversial in its infancy, and voices of opposition nearly led to premature abortion of the project, and much of the objection was voiced due to the doubt of spending a hugh budget in sequencing what was thought to be useless, i.e., for a major proportion (98%) of "junk DNA" at that time. However, recently it was shown that deleting a non-coding region of what was thought to be "junk DNA" could lead to narrowing of arteries in mice [3] so that might offer an explanation that there are more to the story of stroke risk factors than merely studying APOA5, and such findings just remind us time and again, often the basic research could laid the foundation that is very essential for discoveries made by clinicians or the translational/applied scientists.

This issue described expression of a mammalian dyslexia candidate gene KIAA0319L in insect cells [4], and such expression could help in the study of the mechanism of learning disability. The expression level could be improved in the future and that might need further refinement using knowledge derived from study of the codon bias and its application on the expression of foreign genes in other organisms such as fungi- yeasts, molds and mushrooms [reviewed in 5].

---

* E-mail: mary-waye@cuhk.edu.hk

In another article of this series illustrated the complexity of studying Parkinson's disease (PD) as well [6]. It was found that the BDNF polymorphism might not affect the risk of developing Parkinson's disease as shown previously in other populations. Inconsistencies with different findings of the role of BDNF in different research groups could be due to the fact that physical training could lead to a neuroprotection effect on PD patients and such environmental factors plays an important role in affecting levels of BDNF, pro-BDNF as well as the receptor TrkB in the brains of these PD subjects [7]. Such difficulties in evaluation of confounding factors needs to be considered before we could attempt to design therapeutics that need to introduced cross the blood brain barrier [8].

A review was included in this issue to discuss the relationship between genetic factors and epigenetic factors related to bipolar disorder [9], also known as manic depression. It is clear that many genes are involved in bipolar disorder and that environment factors could play a role and they interact with the genome *via* epigenetic changes in a complex manner. At the end, it is hoped that we will understand the regulatory regions of the genome that could be triggered by environmental factors. More knowledge could be gained if we could subject our hypothetical candidate genes to innovative tests to confirm their role in the disease or complex traits, in much the same manner as how the Nobel laureates of physiology prize awarded in 2014:- Dr. John M. O'Keefe, Dr. May-Britt Moser and Dr. Edvard I. Moser performed optogenetic experiments and succeed in their discoveries of the navigation function of nerve cells in the brain [10].

REFERENCES

[1] Pennacchio LA, Olivier M, Hubacek JA, Cohen JC, Cox DR, Fruchart JC, Krauss RM, Rubin EM (Oct 2001). "An apolipoprotein influencing triglycerides in humans and mice revealed by comparative sequencing". *Science* 294 (5540): 169–73.

[2] Hong Mei Wen, Ka Sing Wong, Ho Keung Ng, Wing Sze Cheung, Ru Xun Huang, Brian Tomlinson, G. Neil Thomas, and Larry Baum. (2013) Apolipoprotein A5 polymorphism and triglycerides in ischemic cerebrovascular disease. *Journal of Biochemistry and Molecular Biology in the post genomic era*, Vol 2, no. 1:53-60.

[3] Fang, Janet (February 21, 2010), "Junk DNA holds clues to heart disease: Deleting a non-coding region leads to narrowing of arteries in mice", *Nature*, doi:10.1038/news.2010.82

[4] Savanne Holster, Monique M. van Oers, Els C. Roode, Otto W.H. Tsang, Venus S.Y. Yeung, Just M. Vlak, and Mary M.Y. Waye (2013) Expression of the dyslexia candidate gene Kiaa0319-like in insect cells. *Journal of Biochemistry and Molecular Biology in the post genomic era*, Vol 2, no. 1:45-52.

[5] Ruan Ban-Zhan and Xie Bao-Gui (2013) Analysis of Codon Bias and its Application on the Expression of Foreign Genes in Fungi. *Journal of Biochemistry and Molecular Biology in the post genomic era*, Vol 2, no. 13-14.

[6] Qiwen Liao, Peter Y.K. Poon, Anne Y.Y. Chan, Vincent C.T. Mok, and Larry Baum (2013) A Case-Control Study of Association between Brain-Derived Neurotrophic Factor 712G>A Polymorphism and Parkinson's Disease . *Journal of Biochemistry and Molecular Biology in the post genomic era*, Vol 2, no. 2:127-133.

[7] Tuon T, Valvassori SS, Dal Pont GC, Paganini CS, Pozzi BG, Luciano TF, Souza PS, Quevedo J, Souza CT, Pinho RA (2014) Physical training prevents depressive symptoms and a decrease in brain-derived neurotrophic factor in Parkinson's disease. *Brain Res Bull.* 2014 Sep;108:106-12. doi: 10.1016/j.brainresbull.2014.09.006. Epub 2014 Sep 28.

[8] Ming Wai Yeung and Miu Yee Mary Waye (2013) Genetics and Epigenetics of Bipolar Disorder. *Journal of Biochemistry and Molecular Biology in the post genomic era*, Vol 2, no. 2: 71-100.

[9] Geoffrey H. Morris, Tessa N. Campbell, and Francis Y. M. Choy (2013) The Blood-Brain Barrier Roadblock: Strategies for Targeted Drug Delivery. *Journal of Biochemistry and Molecular Biology in the post genomic era*, Vol 2, no. 2:63-70.

[10] Ole Kiehn (2014) The Brain's Navigational Place and Grid Cell System Retrieved 24th Dec., 2014 from http://www.nobelprize.org/nobel_prizes/medicine/laureates/2014/advanced-medicineprize2014.pdf.

In: Biochemistry and Molecular Biology
Editor: Mary Miu Yee Waye
ISBN: 978-1-63482-312-8

Chapter 1

Apolipoprotein A5 Polymorphism and Triglycerides in Ischemic Cerebrovascular Disease

Hong Mei Wen[1,2], Ka Sing Wong[2], Ho Keung Ng[3], Wing Sze Cheung[3], Ru Xun Huang[1], Brian Tomlinson[2], G. Neil Thomas[4] and Larry Baum[5*]

[1]Department of Neurology, the First Affiliated Hospital of Sun Yat-Sen University, Guangzhou, P.R. China
[2]Departments of Medicine and Therapeutics
[3]Anatomical and Cellular Pathology, The Chinese University of Hong Kong, Shatin, Hong Kong
[4]Department of Public Health, Epidemiology, and Biostatistics, University of Birmingham, Birmingham, UK
[5]School of Pharmacy, The Chinese University of Hong Kong, Shatin, Hong Kong

Abstract

Serum lipids are correlated with certain types of cerebrovascular atherosclerosis. The -1131T>C polymorphism in apolipoprotein A5 (APOA5) is associated with triglyceride levels. We investigated associations among this polymorphism, triglycerides, and stroke in 160 cases of acute ischemic stroke or transient ischemic attack and 157 healthy control subjects. Both patients and controls were Chinese. The APOA5 polymorphism was significantly associated with triglyceride levels both in patients and controls. The average triglyceride level was 70% greater in patients and 92% greater in controls with CC than with TT genotypes. Logistic regression indicated that triglyceride level was an independent risk factor for ischemic stroke in this population. However, APOA5

* Correspondence to Larry Baum, School of Pharmacy, The Chinese University of Hong Kong, Shatin, Hong Kong. E-mail: lwbaum@cuhk.edu.hk, Fax: [852] 26035295, Tel: [852] 39436833

genotypes were not significantly associated with strokes or stroke subtypes, though larger studies would be needed to exclude the possibility of a small effect on stroke risk.

Abbreviations

APOA5 = apolipoprotein A5;
APOA1/C3/A4 = apolipoprotein A1 / apolipoprotein C3 / apolipoprotein A4;
ARIC = atherosclerosis risk in communities;
CI = confidence interval;
CT = computed tomography;
DBP = diastolic blood pressure;
DNA = deoxyribonucleic acid;
dNTP = deoxynucleotide triphosphate;
HDL-C = high-density lipoprotein cholesterol;
LDL-C = low-density lipoprotein cholesterol;
MRA = magnetic resonance angiography;
MRI = magnetic resonance imaging;
OR = odds ratio;
PCR = polymerase chain reaction;
SBP = systolic blood pressure;
SD = standard deviation;
SNP = single nucleotide polymorphism;
TC = total cholesterol;
TCD = transcranial Doppler;
TG = triglyceride;
TIA = transient ischemic attack;
VLDL = very low-density lipoprotein

Introduction

Stroke is one of the most common causes of mortality and morbidity in both developed and developing countries [1]. Although classical risk factors for stroke, such as hypertension and diabetes mellitus, have been established, there is accumulating evidence that genetic elements also contribute to the pathogenesis of cerebrovascular disease [2]. Genes with an established or potential role in atherosclerotic vascular changes appear to be probable candidates, since atherosclerosis has been proposed to be the major underlying condition responsible for acute occlusive cerebrovascular events [3]. Epidemiological studies have suggested that hypertriglyceridemia is an independent risk factor for atherosclerosis [4]. Some but not all studies have reported a positive association between triglyceride levels and ischemic stroke [5;6]. Thus, genetic factors affecting triglyceride levels are possible risk factors for ischemic stroke. For example, the lipoprotein lipase Ser447Ter polymorphism, which is associated with triglyceride level, has been found to affect risk of atherothrombotic cerebral infarction [7].

Pennacchio et al. [8] identified APOA5, an apolipoprotein gene near the APOA1/C3/A4 gene cluster on human chromosome 11q23. Deletion of the gene in knockout mice raised triglyceride levels by a factor of four, while overexpression of APOA5 in transgenic mice lowered triglyceride levels to one-third of normal [8]. Three single nucleotide polymorphisms (SNPs), located 5' to the gene (-1131T>C), in intron 3 (IVS3+476G>A), and in the 3' untranslated region (1259T>C), were associated with triglyceride levels in humans [8]. The -1131T>C polymorphism of APOA5 has been associated with serum triglycerides in populations of different ethnicities [9;10]. The frequency of the rare allele of APOA5 is higher in Chinese than in Hispanic, European or Caucasian populations [8;9;11;12]. The present study was carried out in order to explore the association of this SNP with serum triglyceride levels and the relationship of APOA5 genotype and triglycerides with ischemic stroke.

METHODS

Patients

One hundred and sixty consecutive Chinese patients with acute ischemic stroke or transient ischemic attack (TIA) from January to June 2002 were recruited following admission to the acute stroke unit of a general regional hospital, Prince of Wales Hospital, Shatin, Hong Kong. Research with humans was carried out according to the principles of the Declaration of Helsinki; informed consent was obtained, and the hospital's institutional review board approved the study. All patients in our study were hospitalized because of acute cerebral infarction within seven days of symptom onset. Brain computed tomography (CT) was done on all patients within 24 hours of admission. Patients with past history or CT feature of intracerebral hemorrhage were excluded. Magnetic resonance imaging, including T1, T2-weighted and magnetic resonance angiography (MRA) was performed for most of the patients (90%), except those who were contraindicated for MRI because of pace maker *in-situ*, refusal, claustrophobia, or unstable medical condition. Ultrasound carotid duplex, extracranial and transcranial Doppler (TCD) ultrasound, and electrocardiography were performed routinely for cerebral infarction patients at this hospital.

All patients underwent a detailed clinical analysis. Age, gender and vascular risk factors, including hypertension, diabetes mellitus, smoking habit and history of hyperlipidaemia, were collected during acute admission for all the patients. Subjects were defined as hypertensive if their systolic blood pressure (SBP) was ≥140 mm Hg or diastolic blood pressure (DBP) ≥90 mm Hg on at least two occasions, or if they were receiving blood pressure-lowering medication. Patients were considered diabetic if the fasting plasma glucose was ≥7.8 mmol/L [13].

Lipid profile including triglyceride (TG), high-density lipoprotein cholesterol (HDL-C), and total cholesterol (TC) were measured during hospitalization, most within 3 days after admission. Low-density lipoprotein cholesterol (LDL-C) level was calculated. Levels of TC (enzymatic method), TG (enzymatic method without glycerol blanking), and HDL-C (dextran sulfate-magnesium chloride precipitation) were measured on a Hitachi analyzer.

Etiology

Following evaluation of the clinical and radiological features, we classified the potential vascular etiologies into the following categories:

1. Large artery disease (intracranial large artery disease or extracranial carotid artery disease). Presence of intracranial large artery disease was determined via MRA or TCD. Intracranial large artery disease was considered if the stenosis was at least moderate (≥ 50%) in severity and if the relevant infarct was located within its arterial supply. Signal void was considered as a severe lesion. The criteria of intracranial stenosis by TCD were given elsewhere [14]. Presence of extracranial carotid artery stenosis was screened by duplex ultrasound. Carotid artery stenosis was considered if the stenosis was ≥50% and the relevant infarct was located within its arterial distribution.
2. Small vessel disease. Patients who had no relevant intracranial large artery disease, extracranial carotid artery stenosis, cardiac embolic sources or other miscellaneous causes (such as inflammatory disorders, hematological disorders or recreational drug abuse), and with small infarct (3-15 mm in major diameter present on CT or MRI) located in a subcortical or brain stem area were presumed to have small vessel disease.
3. Cardioembolism was assigned as the etiology for patients who had concurrent presence of arterial fibrillation, mitral or aortic metallic heart valves, acute congestive heart failure, recent (≤6 weeks before stroke) myocardial infarction, sick sinus syndrome, atrial myxoma, or patent foramen ovale. Electrocardiograms were performed on all patients.
4. Other causes were assigned for patients with double or triple pathology or with incomplete or inadequate examinations to draw any conclusion.

Controls

Among people free of symptomatic cerebrovascular diseases who were selected randomly from community centers for the elderly in Shatin, 157 Chinese subjects matched for age and sex volunteered to participate as controls. Venous blood samples from patients and controls were taken in the morning after a 12h fast. All subjects gave written, informed consent to participate in the study, which was approved by the ethics review committee of the hospital.

Genotyping

DNA was extracted from blood. Genotyping was performed as described previously (as SNP3) [8;9]: 200 μM dNTP, 1 unit TaqGold (Applera), 1X TaqGold Buffer, 400 nM AV6F primer (GATTGATTCAAGATGCATTTAGGAC), 400 nM AV6R primer (CCCCAGGAACTGGAGCGAAATT), 1.5 mM $MgCl_2$, and 1 μl DNA were mixed and the volume adjusted to 20 μl with water. PCR was performed at 95°C for 12 min followed by 30 cycles of 94°C for 15 sec, 55°C for 30 sec, and 72°C for 30 sec, followed by 72°C for 3 min. Products were digested with Tru1I or MseI and resolved on 3% agarose gels post-stained with SYBR Gold (Molecular Probes).

Statistics

SPSS version 10.0 and Epi6 (World Health Organization, Geneva, Switzerland) were used for data analysis. Differences in anthropometric and fasting plasma biochemical parameters between patients and controls and among the genotypes were examined using Student's t test and analysis of variance.

Chi-squared analysis was used to compare the genotype distribution between patients and controls and among ischemic stroke subtypes in patients. Comparison of triglyceride levels between patients and controls was performed by Mann-Whitney U test. Kruskal Wallis test was used for comparisons of triglyceride concentration among APOA5 genotypes. To determine the variables correlated with ischemic stroke, we applied multivariate logistic regression. We included all variables that were significant in univariate analysis. These variables included hypertension, diabetes and triglyceride level. Age and gender were also included in the analyses. Because the distribution of triglyceride level was skewed, the natural logarithm of the triglyceride level was introduced into the model.

RESULTS

The study population characteristics are described in Table 1. Stroke patients and controls were matched for age and sex. Both systolic and diastolic blood pressures, as well as frequencies of hypertension and diabetes mellitus, were significantly higher in patients. Triglyceride levels were higher in patients than controls ($p<0.001$).

Table 1. Characteristics of patients and control subjects

	Patients (n=160)	Control (n=157)	p
Age (SD)	70.3 (11.5)	70.0 (2.9)	0.47
Male (%)	83 (51.9)	75 (47.8)	0.46
Hypertension (%)	107 (66.9)	74 (47.1)	<0.001
Diabetes mellitus (%)	49 (30.6)	20 (13.2)	<0.001
Lipid concentration, mmol/L			
TG (SD)	1.75 (0.99)	1.45 (0.95)	<0.001
TC (SD)	5.62 (1.14)	5.42 (0.93)	0.100
HDL-C (SD)	1.35 (0.35)	1.31 (0.36)	0.29
LDL-C (SD)	3.52 (0.99)	3.48 (0.82)	0.72

TC, total cholesterol; HDL-C, high-density lipoprotein cholesterol; LDL-C, low-density lipoprotein cholesterol; TG, triglyceride

Significant differences in triglyceride levels were found among TT, TC and CC genotypes, and the CC genotype was associated with higher triglyceride levels both in patients and in healthy controls (Table 2). The average triglyceride level was 70% greater in patients and 92% greater in controls with CC than with TT genotypes. When patients with allele C are pooled and compared with non-C carriers, allele C carriers are associated with higher triglyceride concentration in controls ($p<0.001$), but not in patients ($p=0.29$).

Table 2. Association of APOA5 genotypes with lipid levels in patients and control subjects

	TT	TC	CC	p
Patients				
TG, mmol/L	1.66 (0.93)	1.69 (0.88)	2.83 (1.60)	0.013
TC, mmol/L	5.63 (1.13)	5.61 (1.08)	5.56 (1.71)	0.98
HDL, mmol/L	1.39 (0.34)	1.32 (0.36)	1.30 (0.35)	0.54
LDL, mmol/L	3.50 (1.00)	3.60 (0.90)	2.97 (1.48)	0.28
Controls				
TG, mmol/L	1.27 (0.96)	1.52 (0.73)	2.44 (0.95)	<0.001
TC, mmol/L	5.52 (0.98)	5.21 (0.87)	5.84 (0.49)	0.031
HDL, mmol/L	1.43 (0.37)	1.19 (0.29)	1.13 (0.39)	<0.001
LDL, mmol/L	3.57 (0.86)	3.32 (0.79)	3.79 (0.56)	0.10

P values for analysis of variance among APOA5 genotypes. Standard deviations in parentheses

This lack of significance may be explained by the similar triglyceride levels in patients with TT and TC genotypes. No significant differences in cholesterol, HDL cholesterol or LDL cholesterol were found among APOA5 genotypes in patients. However, in controls, the total cholesterol ($p=0.031$) and HDL cholesterol levels ($p<0.001$) were different among the genotypes. Comparing controls with CC genotypes with controls having the TT genotype, the average total cholesterol level was 6% higher, and the average HDL level was 21% lower.

Table 3. Frequency of APOA5 genotype in patients and control subjects

	n	TT	TC	CC	P
Control	157	81 (51.6)	66 (42.2)	10 (6.4)	
All patients	160	84 (52.5)	67 (41.9)	9 (5.6)	0.96
Stroke subtype					
Large-vessel disease	58	30 (51.7)	24 (41.4)	4 (6.9)	0.99
Small-vessel disease	49	30 (61.2)	17 (34.7)	2 (4.1)	0.48
Cardioembolism	12	5 (41.7)	6 (50)	1 (8.3)	0.80
Other cause	41	19 (46.3)	20 (48.8)	2 (4.9)	0.73

Chi-square test for comparing genotype distribution between controls and patients as a whole or in subgroups. Percentages in parentheses

Table 3 shows the distribution of APOA5 genotypes according to ischemic stroke subgroups. No significant differences were found between cases and controls in the frequencies of APOA5 genotypes (χ^2=0.086, df=2, p=0.96) or alleles (χ^2=0.46, df=1, p=0.49).

The odds ratio for the association of the C allele with stroke was 0.96, with a 95% confidence interval of 0.66 to 1.38. There were no significant differences in allele frequencies between the sexes in controls (χ^2=0.64, df=2, p=0.73) or in patients (χ^2=0.21, df=2, p=0.90). In patients with large-vessel and cardioembolism etiology, there were trends toward higher frequencies of TC and CC genotypes compared with those with small-vessel disease, but the differences were not significant. No significant differences were observed among ischemic subgroups in triglyceride level (χ^2=0.45, df=3, p=0.72).

Conditional forward logistic regression showed that after adjusting for age and gender, the following variables were associated with ischemic stroke: history of hypertension (OR=1.93, 95% CI 1.19-3.13), diabetes mellitus (OR=2.47, 95% CI 1.36-4.50) and triglyceride level. An increase of 1 natural log unit (2.72x) in triglyceride level was associated with an adjusted OR of 2.22 (95% CI 1.35-3.63).

CONCLUSION

Previous studies showed that APOA5 -1131T>C is significantly associated with plasma triglyceride levels in healthy subjects [9;15], school children [16], and individuals with hypertriglyceridemia [17] or familial combined hyperlipidemia [10]. The purpose of our previous study of this polymorphism was to examine its association with triglyceride levels in healthy subjects [9], while our current study aims to examine its association with the risk of stroke. Our results indicated that the CC genotype was associated with higher triglyceride levels both in stroke patients and control subjects. The association between APOA5 genotype and HDL cholesterol or LDL cholesterol was inconsistent. A number of studies have demonstrated that the rare allele (C) of APOA5 was associated with specific lipoprotein TG and VLDL cholesterol [11] and inversely associated with HDL cholesterol [11;12], while another study [15] failed to find significant difference either in the serum total cholesterol or the LDL and HDL cholesterol among APOA5 genotypes. In the present study, we found the minor allele of APOA5 was inversely associated with HDL cholesterol in controls, which is comparable to previous studies [11;12]. However, no significant association between APOA5 genotype and lipid profiles except for triglyceride was found in stroke patients. This may be due to the relatively small number of patients and lack of statistical power to detect a small effect. Alternatively, it could be due to disease confounders or due to the stroke and subsequent treatment, such as lipid-lowering therapy, affecting the lipid profiles. Larger studies would provide more power to detect small effects, while prospective studies that measure lipid profiles before the stroke may help to clarify the latter possible explanation.

The results suggested that APOA5 variation may be an important factor in determining lipid levels in Chinese. The mechanism by which this polymorphism could affect serum triglyceride and HDL cholesterol levels and propensity to atherosclerosis remains poorly understood. Perhaps –1131C, or a polymorphism in linkage disequilibrium with –1131C [18], influences triglyceride levels by decreasing APOA5 expression or function, and knockout of the APOA5 gene was shown to increase triglyceride levels in mice [8].

Increased plasma triglyceride levels are associated with atherosclerotic heart disease in western countries [19]. However, studies of triglyceride level in relation to stroke or ischemic stroke are not consistent, either reporting a positive association [20] or no association [21]. The atherosclerosis risk in communities (ARIC) study demonstrated a positive association of stroke with triglycerides among women with a history of coronary heart disease, though not in men [22]. In our study, multivariate analysis demonstrated an association between triglyceride levels and ischemic stroke independent of the conventional risk factors of hypertension and diabetes mellitus. However, no relationship was detected between the APOA5 genotype and risk of ischemic stroke as a whole or with the subtypes of cerebral infarction, though an effect of the C allele on stroke risk within the 95% confidence interval of the odds ratio (0.66-1.38) cannot be excluded. This lack of association might be explained by considering that although APOA5 has great influence on plasma triglyceride, other factors such as obesity or lack of exercise are also associated with elevated plasma triglyceride concentrations, but perhaps such other factors increase stroke risk through mechanisms other than triglyceride levels [23]. Alternatively, the number of CC subjects in this study may be too small to show an effect of genotype on stroke risk. Other studies have shown associations between APOA5 genotypes and risk of ischemic stroke, with high-triglyceride genotypes associated with greater risk of stroke, but the reasons for this discrepancy with our results are not known [24-27].

Ischemic stroke is a heterogeneous entity. Atherosclerosis in larger arteries has been related chiefly to lipid levels and hypertension, whereas the atherosclerotic process in smaller arteries has been related to hypertension [28;29]. Therefore, it is necessary to analyze the association between genotype and ischemic subgroups classified by presumed etiology. We found that although there was a trend toward higher C allele frequency among patients with large-artery disease or cardioembolic causes than those patients with small-vessel disease, the difference was not significant in our sample. Triglyceride levels were not different among stroke subtypes.

Several considerations must be given to interpreting the results of our study. First, the size of the studied populations was relatively small, and therefore, our findings need to be confirmed in larger samples and should be tested in different ethnic groups. Furthermore, we did not take into account other factors that might have potential effect on triglyceride levels, such as body mass index, which may affect the robustness of the results.

In summary, APOA5 –1131T>C genotype was associated with plasma triglyceride levels in Chinese, and triglyceride concentration was a risk factor for ischemic stroke in this population. We can exclude a large effect of APOA5 genotype on stroke (one beyond our confidence interval of the odds ratio), and larger studies would be needed to narrow the confidence interval and exclude the possibility of a smaller effect.

REFERENCES

[1] Stegmayr, B; Asplund, K; Kuulasmaa, K; Rajakangas, AM; Thorvaldsen, P; Tuomilehto, J. Stroke incidence and mortality correlated to stroke risk factors in the

WHO MONICA Project. An ecological study of 18 populations. *Stroke*, 1997 28, 1367-1374.

[2] Hassan, A; Markus, HS. Genetics and ischaemic stroke. *Brain*, 2000 123, 1784-1812.

[3] Quandt, CM; Talbert, RL; de los Reyes, RA. Current concepts in clinical therapeutics: ischemic cerebrovascular disease. *Clin. Pharm.*, 1987 6, 292-306.

[4] Hodis, HN; Mack, WJ. Triglyceride-rich lipoproteins and the progression of coronary artery disease. *Curr. Opin. Lipidol.*, 1995 6, 209-214.

[5] Njolstad, I; Arnesen, E; Lund-Larsen, PG. Body height, cardiovascular risk factors, and risk of stroke in middle-aged men and women. A 14-year follow-up of the Finnmark Study. *Circulation*, 1996 94, 2877-2882.

[6] Rizos, E; Mikhailidis, DP. Are high density lipoprotein (HDL) and triglyceride levels relevant in stroke prevention? *Cardiovasc. Res.*, 2001 52, 199-207.

[7] Shimo-Nakanishi, Y; Urabe, T; Hattori, N; Watanabe, Y; Nagao, T; Yokochi, M; Hamamoto, M; Mizuno, Y. Polymorphism of the lipoprotein lipase gene and risk of atherothrombotic cerebral infarction in the Japanese. *Stroke*, 2001 32, 1481-1486.

[8] Pennacchio, LA; Olivier, M; Hubacek, JA; Cohen, JC; Cox, DR; Fruchart, JC; Krauss, RM; Rubin, EM. An apolipoprotein influencing triglycerides in humans and mice revealed by comparative sequencing. *Science*, 2001 294, 169-173.

[9] Baum, L; Tomlinson, B; Thomas, GN. APOA5-1131T>C polymorphism is associated with triglyceride levels in Chinese men. *Clin. Genet.*, 2003 63, 377-379.

[10] Ribalta, J; Figuera, L; Fernandez-Ballart, J; Vilela, E; Castro, CM; Masana, L; Joven, J. Newly identified apolipoprotein AV gene predisposes to high plasma triglycerides in familial combined hyperlipidemia. *Clin. Chem.*, 2002 48, 1597-1600.

[11] Aouizerat, BE; Kulkarni, M; Heilbron, D; Drown, D; Raskin, S; Pullinger, CR; Malloy, MJ; Kane, JP. Genetic analysis of a polymorphism in the human apoA-V gene: effect on plasma lipids. *J. Lipid. Res.*, 2003 44, 1167-1173.

[12] Lai, CQ; Tai, ES; Tan, CE; Cutter, J; Chew, SK; Zhu, YP; Adiconis, X; Ordovas, JM. The APOA5 locus is a strong determinant of plasma triglyceride concentrations across ethnic groups in Singapore. *J. Lipid. Res.*, 2003 44, 2365-2373.

[13] WHO. Report of the Expert Committee on the Diagnosis and Classification of Diabetes Mellitus. *Diabetes Care*, 1997 20, 1183-1197.

[14] Wong, KS; Li, H; Chan, YL; Ahuja, A; Lam, WW; Wong, A; Kay, R. Use of transcranial Doppler ultrasound to predict outcome in patients with intracranial large-artery occlusive disease. *Stroke*, 2000 31, 2641-2647.

[15] Nabika, T; Nasreen, S; Kobayashi, S; Masuda, J. The genetic effect of the apoprotein AV gene on the serum triglyceride level in Japanese. *Atherosclerosis*, 2002 165, 201-204.

[16] Endo, K; Yanagi, H; Araki, J; Hirano, C; Yamakawa-Kobayashi, K; Tomura, S. Association found between the promoter region polymorphism in the apolipoprotein A-V gene and the serum triglyceride level in Japanese schoolchildren. *Hum. Genet.*, 2002 111, 570-572.

[17] Horinek, A; Vrablik, M; Ceska, R; Adamkova, V; Poledne, R; Hubacek, JA. T-1131-->C polymorphism within the apolipoprotein AV gene in hypertriglyceridemic individuals. *Atherosclerosis*, 2003 167, 369-370.

[18] Pennacchio, LA; Olivier, M; Hubacek, JA; Krauss, RM; Rubin, EM; Cohen,JC. Two independent apolipoprotein A5 haplotypes influence human plasma triglyceride levels. *Hum. Mol. Genet.*, 2002 11, 3031-3038.

[19] Malloy, MJ; Kane, JP. A risk factor for atherosclerosis: triglyceride-rich lipoproteins. *Adv. Intern. Med.*, 2001 47, 111-136.

[20] Lindenstrom, E; Boysen, G; Nyboe, J. Influence of total cholesterol, high density lipoprotein cholesterol, and triglycerides on risk of cerebrovascular disease: the Copenhagen City Heart Study. *BMJ*, 1994 309, 11-15.

[21] Wannamethee, SG; Shaper, AG; Ebrahim, S. HDL-Cholesterol, total cholesterol, and the risk of stroke in middle-aged British men. *Stroke*, 2000 31, 1882-1888.

[22] Shahar, E; Chambless, LE; Rosamond, WD; Boland, LL; Ballantyne, CM; McGovern, PG; Sharrett, AR. Plasma lipid profile and incident ischemic stroke: the Atherosclerosis Risk in Communities (ARIC) study. *Stroke*, 2003 34, 623-631.

[23] Connelly, PW; Petrasovits, A; Stachenko, S; MacLean, DR; Little, JA; Chockalingam, A. Prevalence of high plasma triglyceride combined with low HDL-C levels and its association with smoking, hypertension, obesity, diabetes, sedentariness and LDL-C levels in the Canadian population. Canadian Heart Health Surveys Research Group. *Can. J. Cardiol.*, 1999 15, 428-433.

[24] Havasi, V; Szolnoki, Z; Talian, G; Bene, J; Komlosi, K; Maasz, A; Somogyvari, F; Kondacs, A; Szabo, M; Fodor, L; Bodor, A; Melegh, B. Apolipoprotein A5 gene promoter region T-1131C polymorphism associates with elevated circulating triglyceride levels and confers susceptibility for development of ischemic stroke. *J. Mol. Neurosci.*, 2006 29, 177-183.

[25] Li, J; Xu, HW; Zhu, XY. [Association of APOA5 gene polymorphism with levels of lipids and atherosclerotic cerebral infarction in Chinese]. *Zhonghua Yi. Xue. Yi. Chuan Xue. Za Zhi.*, 2007 24, 576-578.

[26] Li, X; Su, D; Zhang, X; Zhang, C. Association of apolipoprotein A5 gene promoter region -1131T>C with risk of stroke in Han Chinese. *Eur. J. Intern. Med.*, 2011 22, 99-102.

[27] Tai, ES; Ordovas, JM. Clinical significance of apolipoprotein A5. *Curr. Opin. Lipidol.*, 2008 19, 349-354.

[28] Caplan, LR; Gorelick, PB; Hier, DB. Race, sex and occlusive cerebrovascular disease: a review. *Stroke*, 1986 17, 648-655.

[29] Mathew, NT; Davis, D; Meyer, JS; Chandar, K. Hyperlipoproteinemia in occlusive cerebrovascular disease. *JAMA*, 1975 232, 262-266.

In: Biochemistry and Molecular Biology
Editor: Mary Miu Yee Waye
ISBN: 978-1-63482-312-8

Chapter 2

EXPRESSION OF THE DYSLEXIA CANDIDATE GENE KIAA0319-LIKE IN INSECT CELLS

Savanne Holster[1], Monique M. van Oers[1], Els C. Roode[1], Otto W. H. Tsang[2], Venus S. Y. Yeung[2], Just M. Vlak[1] and Mary M. Y. Waye[2*]

[1]Laboratory of Virology, Wageningen University, the Netherlands
[2]School of Biomedical Sciences, Croucher Laboratory for Human Genomics, The Chinese University of Hong Kong, Shatin, N.T. Hong Kong

ABSTRACT

The human kiaa0319-like gene is one of the candidate genes for developmental dyslexia, but the exact function of the encoded KIAA0319L (KL) protein is not known. To allow functional analysis a purified, biologically active KL protein is required. The *kiaa0319-like* gene was expressed in insect cells using the baculovirus expression system. To optimize the expression of the *kiaa0319-like* gene and to be able to purify the KL protein, several approaches were used. Two different recombinant baculoviruses were made, one with the full length coding sequence of KL and one that lacked the transmembrane domain to facilitate purification.. Versions in which the *kiaa0319L* sequences were cloned downstream of the honeybee melittin signal sequence were also made. All four constructs contained a C-terminal influenza hemagglutinin (HA)-tag. Sf9 insect cells infected with these recombinant baculoviruses produced the KL protein, as demonstrated by Western blot analysis using either the HA-antibody or KL-specific polyclonal serum.

Keywords: Dyslexia, baculovirus expression system, *kiaa0319*, *kiaa0319-like*

* Corresponding author: Rm 324A, Lo Kwee-Seong Integrated Biomedical Sciences Building, School of Biomedical Sciences, The Chinese University of Hong Kong, Shatin, N.T. Hong Kong, Tel: (852) 39436874, Fax: (852) 2603-5123, E-mail: mary-waye@cuhk.edu.hk

Main Text

Five to ten percent of school-age children are affected by the disorder developmental dyslexia (Habib, 2000), which is characterized by a difficulty to read and write. Children with dyslexia have more difficulties in learning the alphabetic and logographic principles compared to non-affected children of similar age. Therefore, these children do not learn the skills to read and write as readily as non-affected children (Gabrieli, 2009). Developmental dyslexia is not due to a deficit in general intelligence or a lack of motivation nor to an inappropriate educational environment, but due to as yet unknown environmental effects and/or genetic alterations in brain morphology and function. Dyslexia is normally diagnosed when children are about eight years old, and some are diagnosed even at an older age. Although several tests are available to check if a child has dyslexia it is often difficult to distinguish between children, who have the disorder and children who do not have dyslexia. Many children are therefore misdiagnosed. There is a need for a diagnostic test that would allow proper diagnosis and that can be applied at an early age. Such a test could be based on biological markers and should ideally include all dyslexia related genes.

So far nine quantitative trait loci (DYX1-9) have been found with a connection to dyslexia (Williams and O'Donovan, 2006). A number of genes on these loci have been proposed as susceptibility genes, including *dyx1c1* for the DYX1 locus (Bates et al., 2010; Bellini et al., 2005; Lim et al., 2011; Taipale et al., 2003; Zhang et al., 2012), *kiaa0319* and *dcdc2* for the DYX2 locus (Cope, 2005; Meng et al., 2005; Paracchini et al., 2006; Schumacher et al., 2007), *robo1* for the DYX5 (Hannula-Jouppi et al., 2005) and *mrpl19* and *c2orf3* for the DYX3 locus (Anthoni et al., 2007). A marker located on chromosome 4q32.1 that has a possible trans-regulatory effect on the gene for SLC2A3 (solute carrier 2 A3) on chromosome 12 (Roeske et al., 2011), has also been reported to have association with dyslexia. DGKI in chromosome 7q33 has also been associated with dyslexia in the Finnish and German human populations (Matsson et al., 2011).

Proteins encoded by some of these genes are thought to participate in brain developmental processes including neuron migration and axon guidance, and deficiencies of these proteins may lead to abnormalities in brain development, which may affect processes critical for learning (Galaburda et al., 2006). Among these genes *kiaa0319* is very well characterized. The encoded KIAA0319 protein is both N- and O-glycosylated and is localized in the plasma membrane as a dimer (Velayos-Baeza et al., 2008). The gene is expressed in the developing cerebral neocortex, when neuronal migration takes place (Peschansky et al., 2010). When *kiaa0319* expression is artificially suppressed by RNAi, neurons are not able to migrate, because they do not associate with glial fibers (Paracchini et al., 2006).

Recently, several novel genes have been suggested to be associated with developmental dyslexia, including a homologue of *kiaa0319*, which is addressed as *kiaa0319-like* (*kl*) (Couto et al., 2008), and which is located on human chromosome 1p34. The proteins encoded by *kiaaa0319* and *kl* have a similarity of 61%. The *kl* gene has been identified as a candidate for developmental dyslexia (Couto et al., 2008), but only with a modestly significant association. The *kl* gene is expressed in various human tissues, including the adult and fetal brain (Poon et al., 2011a). The expression of *kl* is most abundant in the cerebellum, where the KL protein was detected in the cytoplasm, the periphery of the nucleus and also in the nucleus (Poon et al., 2011a). In a later report, a more precise localization of the protein in the cytoplasmic granules of the cortex of the human brain was deciphered using immunohistochemistry. The results confirmed expression of *kl* in brain regions that are involved in reading performance, suggesting that malfunctioning of the *kl* gene may lead to reading disabilities (Poon et al., 2011c). Using a yeast-2-hybrid screen and co-immunoprecipitation assays, it was demonstrated that the KL protein has physical interactions with Nogo Receptor 1 (NgR1), an axon guidance receptor (Poon et al., 2011b). This receptor is a key protein in axon elongation and growth cone collapse in brain development (Hasegawa et al., 2004). Interaction between KL and NgR1 is found in cytoplasmic granules of cortical neurons in the human brain cortex (Poon et al., 2011b).

To further analyze the function of the KL protein, purified, biologically functional KL protein would be a useful tool. Therefore, we aimed in this study at producing the KL protein in insect cells using the baculovirus expression system. By performing computational analyses a single transmembrane domain (TM), located in exon 19, has been predicted for the KLAA0319 protein (Velayos-Baeza et al., 2007). Based on amino acid sequence alignments and hydrophobicity profiles, a TM is also predicted for KL (Figure 1). In the current study constructs were made with and without the predicted TM domain. The constructs that lacked the TM also lacked the C-terminal tail (123 codons deleted in total). Because it is not known if this deletion will seriously alter the function of the KL protein or even lead to a non-functional protein, a construct containing the full length coding sequence was also made (Figure 1). All constructs were generated with a C-terminal influenza hemagglutinin (HA) tag for easy detection and to allow future purification. Apart from the TM mentioned above, a hydrophobic region is also predicted between amino acid 30 and 50 (Figure 1), but whether this region can serve a signal peptide in insect cells is not known. Therefore constructs were also made, in which the KL ORF was cloned downstream of the honeybee melittin (HBM) signal sequence (Tessier et al., 1991). We and others have used this HBM signal before to efficiently direct mammalian proteins to the cell surface (e.g., Kaba et al., 2004).

To make the kl and klΔTM constructs (Figure 1) a polymerase chain reaction (PCR) was performed with Phusion High Fidelity DNA polymerase (Finnzymes). The upstream primer kl-F-new was used in combination with primer kl-HA-NotI-R for the full length construct and with kl-deltaTM-R to generate klΔTM (see Table 1 for primer sequences). Plasmid KLgfpC1 served as template (Poon et al, 2011b).

Table 1. List of PCR and sequence primers*

kl-F new	5'- TAGGTCGACTAAATATGGAGAAGAGGCTGGGAGT -3'
kl-HA-NotI-R	5'- TAGGCGGCCGCTCTAGACTACGCATAGTC CGGCAC GTCGTACGGATACAGGATCTCCTCCCGCGGGC -3'
kl-deltaTM-R	5'- TAGGCGGCCGCTCTAGACTACGCATAGTCCGGCACGTC GTACGGATAACAGTTGCTGTCTCCATC-3'
Fw800kl	5'-CGCCCAGCACTCAACAAGTAA-3'
Fw1600kl	5'-AGCTATGAGTGGTCACTCAGC-3'

*Sequences that introduce restriction sites are underlined.

The upstream primer kl-F-new added a *Sal*I restriction site followed by five extra nucleotides (nt) before the ATG start codon: The nt AAT were added directly upstream of the ATG start codon to keep in line with common rules of baculovirus translational start regions

(Chang et al., 1999). The AAT triplet was preceded by two extra nt to avoid frame shifting upon cloning downstream of the melittin signal sequence. The reverse primers introduced the HA-tag followed by a stop codon and restriction sites specific for the enzymes *Xba*I and *Not*I. The *Not*I site was eliminated in the final cloning step. The PCR products for kl (3,208 bp) and klΔTM (2,839 bp) were first cloned in the pJET1.2 vector (Clontech) and verified by sequence analysis at Macrogen, using the standard pJET1.2 forward sequencing primers (Clontech), and the specific internal primers Fw800kl and Fw1600kl (Table 1).

Both sequences were recloned downstream of the baculovirus polyhedrin promoter as *Sal*I-*Xba*I fragments into pFastBacDual (pFBD; Invitrogen) and pFastBac-Melittin transfer vectors (pFBM), described by (Kaba et al., 2004). This procedure resulted in the four transfer vectors pFBD-kl, pFBD-klΔTM, pHBM-kl, and pHBM-klΔTM.

Bacmid technology was applied to generate recombinant baculoviruses by transposition in *E. coli* (Luckow, 1993) following the Bac-to-Bac protocol (Invitrogen). A modified bacmid (Metz et al., 2011) was used that lacked the *p10* promoter and open reading frame (ORF) and the chitinase and cathepsin promoters and large parts of these ORFs (AcBac10ΔccΔp10). Recombination was performed in DH10ß *E. coli* cells from which the genomic Tn7 insertion site was removed to avoid false positives upon transposition (Airenne et al., 2003). The recombinant bacmid clones were analyzed by PCR to verify the absence of empty bacmids and used to transfect Sf9 insect cells using Cellfectin®II Reagent (Invitrogen).

Recombinant budded viruses were harvested 4 days post transfection and used to produce high titer virus stocks in fresh Sf9 cells. Sf9 cells were routinely grown in monolayer cultures in Sf900II medium supplemented with fetal bovine serum (Hyclone) and gentamycin (10 mg/ml).

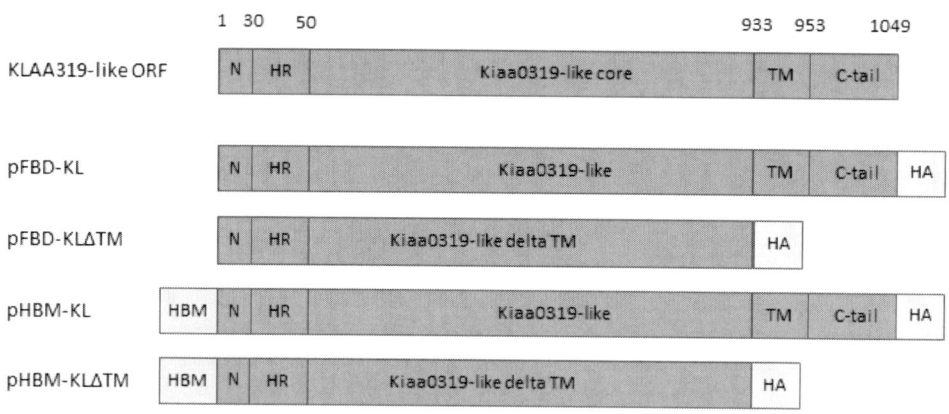

Figure 1. Diagram of the constructs made to express the human Kiaaa0319-like (*kl*) open reading frame (A) in insect cells using baculovirus vectors. Two sequences were amplified by PCR: 1) the full length ORF with an N-terminal domain (N), followed by a hydrophobic domain (HD), and the remainder of the ORF, and 2) a truncated version that lacked a predicted transmembrane (TM) region and the downstream C-terminal tail. The primers used in this PCR were designed to add restriction sites, C-terminal hemagglutinin (HA)-tags and new stop codons. The verified PCR products were recloned downstream of the polyhedrin promoter of the baculovirus transfer vectors pFastBacDUAL (pFBD) (B and D), or downstream of the honey bee melittin signal (HMB) sequence in pFastBac-Melittin (pHBM) (C and E). The plasmid constructs were used to make recombinant baculoviruses via the Bac-to-Bac system (Invitrogen).

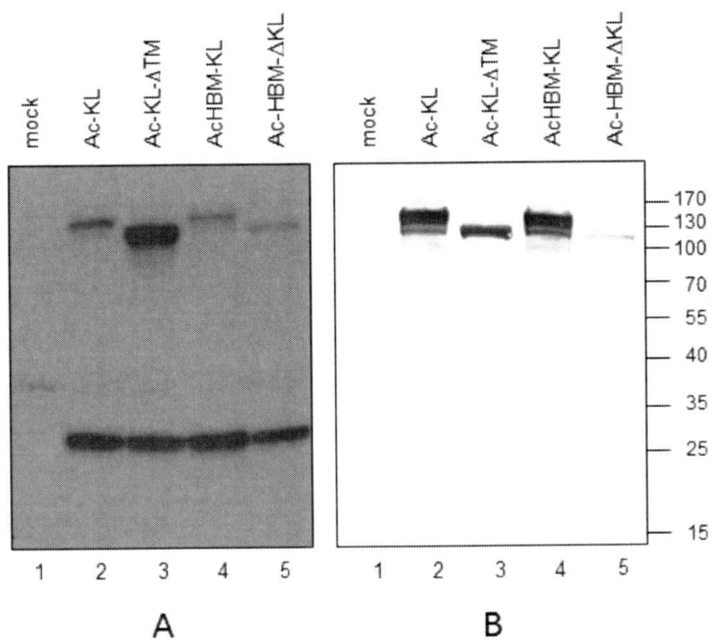

Figure 2. KL protein production in insect cells. Sf9 insect cells were either mock infected (lanes 1), or infected with Ac-KL (2), Ac-KLΔTM (3), AcHBM-KL (4) or AcHBM-KLΔTM (5) and processed at 48 h post infection for Western blotting. The blots were developed with a KL-specific antiserum (A) or with anti-HA antibody (B). Marker sizes are indicated in kDa on the right.

To test whether the inserted kl sequences were expressed, flasks with 2 x 10^6 Sf9 cells were infected with a multiplicity of infection of 10 TCID50 units/cell with the various recombinant viruses. At 48 h post infection the infected cells were harvested and washed with and suspended in phosphate buffered saline (PBS) containing protease inhibitor cocktail (Complete; Roche Applied Science). Protein samples (equivalent to 2 x 10^5 cells per lane) were analyzed in Any kD™ Mini-PROTEAN® TGX™ precast gels (BIORAD) and subsequently transferred by semi-dry blotting to Immobilon-P membrane for Western blot analysis (Figure 2). Immunodetection was performed with KL-specific rabbit polyclonal antiserum purified over a Protein A affinity column and diluted 1:1000 (Poon et al., 2011b). The second antibody used for this assay was 1: 2000 diluted goat-anti-rabbit IgG- horse radish peroxidase conjugated (Dako) (Figure 2A). A duplicate blot was incubated with rat anti-HA-monoclonal antibody (1:1000 diluted, Roche) and with goat-anti-rat IgG coupled to alkaline phosphatase (1:2000, Sigma) as second antibody. This blot was developed with NBT-BCIP (Roche) (Figure 2B).

The predicted size for the KIAA0319 protein (without HA tag) is 115.8 kDa, but the protein has been reported to be N-glycosylated in humans (Velayos-Baeza et al., 2008). In a study in which the KL protein present in adult mouse brain was analyzed, KL-specific polypeptides of 140, 160 and 190 kDa were detected on Western blots (Poon et al., 2011c), which may be to glycosylation in combination with differential processing. In our analysis we see for the full length construct, Ac-KL, a protein around 140 kDa reacting with the KL-specific antiserum (Figure 2A, lane 2). This protein is also detected with the HA-specific antibody (Figure 2B, lane 2), as well as a less abundant polypeptide around 130 kDa, which may represent a not fully glycosylated form of KL. For the construct without the TMR domain Ac-KLΔTM (Figure 2, lanes 3), a protein is detected around 130 kDa. The relative amount of the protein with this construct detected varies whether we use the KL or HA antibody, suggesting that part of the final product may not contain the HA-tag anymore.

With the KL-specific antiserum we also detected a polypeptide of approximately 28 kDa, which appears to be a processed or degraded form (despite the presence of protease inhibitors and the use of a chitinase and cathepsin deficient baculovirus vector). As this 28 kDa polypeptide is detected in cells infected with either the full length or the C-terminally truncated version, we conclude that this polypeptide does not contain the TM and C-terminal tail. As the HA antibody is not able to detect this 28 kDa form in either construct, the C-terminal HA-tag is not present either, and the polypeptide is most likely generated from a more N-terminally located sequence.

When the HBM signal sequence is present upstream of the KL protein, similar protein sizes are seen as with the native N-terminus (Figure 2, lanes 4 and 5). With a HBM signal sequence and without a TM region considerably less KL protein is detected inside the cells. To analyze whether this was the consequence of KL secretion from the cells, we tried to locate KL in the culture medium, but so far no KL protein was detected with any of the four constructs, irrespective of the presence or absence of a TM region (data not shown). We hypothesize that the N-terminal hydrophobic domain (aa 30-50, Figure 1) may function both as a non-removed signal peptide and as a membrane anchor sequence.

In this study we aimed at the expression of the dyslexia-related *kiaa0319-like* gene using the baculovirus-insect cell expression system. This system was chosen to allow high levels of expression and posttranslational modifications, a prerequisite for the production of functional eukaryotic proteins. We were able to produce the KL protein in a full length form and in a C-terminally truncated form, using two different expression vectors. As both forms also contained a HA-tag as was confirmed by western blot analysis, the next step will be the purification of the recombinant proteins based on the HA-tag and to use the purified protein to search for specific ligands for KL in human brain tissue.

ACKNOWLEDGMENTS

Paul Sondermeijer (MSD Animal Health, Boxmeer the Netherlands) is acknowledged for the pFastBac-Melittin vector. We are grateful to Kari Airenne (University of Eastern Finland, Kuopio) for the DH10ß cells with the Tn7 mutation. Shirisha Velpula is acknowledged for her assistance as an MSc student at Wageningen University. This project was supported in part by joint research grant No. 600.649.000.10N004 from the Netherlands Organization for Scientific Research (NWO) and the Research Grants Council of Hong Kong (RGC).

REFERENCES

Airenne, K.J., Peltomaa, E., Hytonen, V.P., Laitinen, O.H., Yla-Herttuala, S., 2003. Improved generation of recombinant baculovirus genomes in *Escherichia coli*. *Nucleic. Acids Res.* 31, e101.

Anthoni, H., Zucchelli, M., Matsson, H., Muller-Myhsok, B., Fransson, I., Schumacher, J., Massinen, S., Onkamo, P., Warnke, A., Griesemann, H., Hoffmann, P., Nopola-Hemmi, J., Lyytinen, H., Schulte-Koerne, G., Kere, J., Nothen, M.M., Peyrard-Janvid, M., 2007. A locus on 2p12 containing the co-regulated MRPL19 and C2ORF3 genes is associated to dyslexia. *Hum. Mol. Genet.* 16, 667-677.

Bates, T.C., Lind, P.A., Luciano, M., Montgomery, G.W., Martin, N.G., Wright, M.J., 2010. Dyslexia and DYX1C1: deficits in reading and spelling associated with a missense mutation. *Mol. Psychiatry* 15, 1190-1196.

Bellini, G., Bravaccio, C., Calamoneri, F., Cocuzza, M.D., Fiorillo, P., Gagliano, A., Mazzone, D., del Giudice, E.M., Scuccimarra, G., Militerni, R., Pascotto, A., 2005. No evidence for association between dyslexia and DYX1C1 functional variants in a group of children and adolescents from Southern Italy. *J. Mol. Neurosci.* 27, 311-314.

Chang, M., Kuzio, J., Blissard, G., 1999. Modulation of translational efficiency by contextual nucleotides flanking a baculovirus initiator AUG codon. *Virology* 259, 369-383.

Cope, 2005. Strong evidence that KIAA0319 on chromosome 6p is a susceptibility gene for developmental dyslexia (vol 76, pg 581, 2005). *Am. J. Hum. Genet.* 77, 898-898.

Couto, J.M., Gomez, L., Wigg, K., Cate-Carter, T., Archibald, J., Anderson, B., Tannock, R., Kerr, E.N., Lovett, M.W., Humphries, T., Barr, C.L., 2008. The KIAA0319-like (KIAA0319L) gene on chromosome 1p34 as a candidate for reading disabilities. *J. Neurogenet.* 22, 295-313.

Gabrieli, J.D., 2009. Dyslexia: a new synergy between education and cognitive neuroscience. *Science* 325, 280-283.

Galaburda, A.M., LoTurco, J., Ramus, F., Fitch, R.H., Rosen, G.D., 2006. From genes to behavior in developmental dyslexia. *Nat. Neurosci.* 9, 1213-1217.

Habib, M., 2000. The neurological basis of developmental dyslexia: an overview and working hypothesis. *Brain* 123 2373-2399.

Hannula-Jouppi, K., Kaminen-Ahola, N., Taipale, M., Eklund, R., Nopola-Hemmi, J., Kaariainen, H., Kere, J., 2005. The axon guidance receptor gene ROBO1 is a candidate gene for developmental dyslexia. *Plos. Genet.* 1, 467-474.

Hasegawa, Y., Fujitani, M., Hata, K., Tohyama, M., Yamagishi, S., Yamashita, T., 2004. Promotion of axon regeneration by myelin-associated glycoprotein and Nogo through divergent signals downstream of Gi/G. *J. Neurosci.* 24, 6826-6832.

Kaba, S.A., Salcedo, A.M., Wafula, P.O., Vlak, J.M., van Oers, M.M., 2004. Development of a chitinase and v-cathepsin negative bacmid for improved integrity of secreted recombinant proteins. *J. Virol. Methods* 122, 113-118.

Lim, C.K.P., Ho, C.S.H., Chou, C.H.N., Waye, M.M.Y., 2011. Association of the rs3743205 variant of DYX1C1 with dyslexia in Chinese children. *Behav. Brain Funct.* 7, 6.

Luckow, V.A., 1993. Baculovirus systems for the expression of human gene products. *Curr. Opin. Biotechnol.* 4, 564-572.

Matsson, H., Tammimies, K., Zucchelli, M., Anthoni, H., Onkamo, P., Nopola-Hemmi, J., Lyytinen, H., Leppanen, P.H.T., Neuhoff, N., Warnke, A., Schulte-Korne, G., Schumacher, J., Nothen, M.M., Kere, J., Peyrard-Janvid, M., 2011. SNP variations in the 7q33 region containing DGKI are associated with dyslexia in the Finnish and German populations. *Behavior. Genetics* 41, 134-140.

Meng, H.Y., Smith, S.D., Hager, K., Held, M., Liu, J., Olson, R.K., Pennington, B.F., DeFries, J.C., Gelernter, J., O'Reilly-Pol, T., Somlo, S., Skudlarski, P., Shaywitz, S.E.,

Shaywitz, B.A., Marchione, K., Wang, Y., Paramasivam, M., LoTurco, J.J., Page, G.P., Gruen, J.R., 2005. DCDC2 is associated with reading disability and modulates neuronal development in the brain. *Proc. Natl. Acad. Sci. U S A* 102, 17053-17058.

Metz, S.W., Geertsema, C., Martina, B.E., Andrade, P., Heldens, J.G., van Oers, M.M., Goldbach, R.W., Vlak, J.M., Pijlman, G.P., 2011. Functional processing and secretion of Chikungunya virus E1 and E2 glycoproteins in insect cells. *Virol J.* 8, 353.

Paracchini, S., Thomas, A., Castro, S., Lai, C., Paramasivam, M., Wang, Y., Keating, B.J., Taylor, J.M., Hacking, D.F., Scerri, T., Francks, C., Richardson, A.J., Wade-Martins, R., Stein, J.F., Knight, J.C., Copp, A.J., Loturco, J., Monaco, A.P., 2006. The chromosome 6p22 haplotype associated with dyslexia reduces the expression of KIAA0319, a novel gene involved in neuronal migration. *Hum. Mol. Genet.* 15, 1659-1666.

Peschansky, V.J., Burbridge, T.J., Volz, A.J., Fiondella, C., Wissner-Gross, Z., Galaburda, A.M., Lo Turco, J.J., Rosen, G.D., 2010. The effect of variation in expression of the candidate dyslexia susceptibility gene homolog Kiaa0319 on neuronal migration and dendritic morphology in the rat. *Cereb. Cortex* 20, 884-897.

Poon, M.W., Chan, H.L., Lim, K.P., Waye, M.M.Y., 2011a. The dyslexia candidate gene kiaa03191 encodes N-glycosylated isoforms that form homo-dimers. *J. Biochem. Mol. Biol. Post. Genomic. Era* 1, 4.

Poon, M.W., Tsang, W.H., Chan, S.O., Li, H.M., Ng, H.K., Waye, M.M., 2011b. Dyslexia-associated kiaa0319-like protein interacts with axon guidance receptor nogo receptor 1. *Cell. Mol. Neurobiol.* 31, 27-35.

Poon, M.W., Tsang, W.H., Waye, M.M., Chan, S.O., 2011c. Distribution of Kiaa0319-like immunoreactivity in the adult mouse brain - a novel protein encoded by the putative dyslexia susceptibility gene KIAA0319-like. *Histol. Histopathol.* 26, 953-963.

Roeske, D., Ludwig, K.U., Neuhoff, N., Becker, J., Bartling, J., Bruder, J., Brockschmidt, F.F., Warnke, A., Remschmidt, H., Hoffmann, P., Muller-Myhsok, B., Nothen, M.M., Schulte-Korne, G., 2011. First genome-wide association scan on neurophysiological endophenotypes points to trans-regulation effects on SLC2A3 in dyslexic children. *Mol. Psychiatry* 16, 97-107.

Schumacher, J., Hoffmann, P., Schmal, C., Schulte-Korne, G., Nothen, M.M., 2007. Genetics of dyslexia: the evolving landscape. *J. Med. Genet.* 44, 289-297.

Taipale, M., Kaminen, N., Nopola-Hemmi, J., Haltia, T., Myllyluoma, B., Lyytinen, H., Muller, K., Kaaranen, M., Lindsberg, P.J., Hannula-Jouppi, K., Kere, J., 2003. A candidate gene for developmental dyslexia encodes a nuclear tetratricopeptide repeat domain protein dynamically regulated in brain. *Proc. Natl. Acad. Sci. U S A* 100, 11553-11558.

Tessier, D.C., Thomas, D.Y., Khouri, H.E., Laliberte, F., Vernet, T., 1991. Enhanced secretion from insect cells of a foreign protein fused to the honeybee melittin signal peptide. *Gene.* 98, 177-183.

Velayos-Baeza, A., Toma, C., da Roza, S., Paracchini, S., Monaco, A.P., 2007. Alternative splicing in the dyslexia-associated gene KIAA0319. *Mamm. Genome* 18, 627-634.

Velayos-Baeza, A., Toma, C., Paracchini, S., Monaco, A.P., 2008. The dyslexia-associated gene KIAA0319 encodes highly N- and O-glycosylated plasma membrane and secreted isoforms. *Hum. Mol. Genet.* 17, 859-871.

Williams, J., O'Donovan, M.C., 2006. The genetics of developmental dyslexia. *Eur. J. Hum. Genet.* 14, 681-689.

Zhang, Y.P., Li, J., Tardif, T., Burmeister, M., Villafuerte, S.M., McBride-Chang, C., Li, H., Shi, B.J., Liang, W.L., Zhang, Z.X., Shu, H., 2012. Association of the DYX1C1 Dyslexia Susceptibility Gene with Orthography in the Chinese Population. *Plos One* 7.

In: Biochemistry and Molecular Biology
Editor: Mary Miu Yee Waye

ISBN: 978-1-63482-312-8

Chapter 3

The Lipoprotein Lipase 447Ter Variant Is Not Associated with Age-Related Macular Degeneration in Chinese

Larry Baum[1], **Wai-Man Chan**[2], **Kai-On Chu**[3], **Fion T. C. Lau**[2], **Alvin K. H. Kwok**[2], **Dennis S. C. Lam**[2] **and Chi P. Pang**[2]*

[1]School of Pharmacy, [2]Department of Ophthalmology and Visual Sciences
[3]Department of Obstetrics and Gynecology,
The Chinese University of Hong Kong, Hong Kong, China

Abstract

Background: Lipoprotein lipase (LPL) promotes the binding and transfer of apolipoprotein E (ApoE) lipoprotein particle components to cells. Age-related macular degeneration (AMD), characterized by drusen, containing lipids, is reportedly associated with a lower frequency of the ApoE ε4 allele than are control subjects. Because of the functional association of LPL and ApoE, we sought to examine whether an LPL polymorphism affects AMD risk. The common LPL alteration Ser447Ter decreases triglyceride levels and risk of cardiovascular disease. *Methods:* We genotyped this LPL polymorphism in 140 unrelated Hong Kong Chinese AMD patients and 100 elderly normal controls using polymerase chain reaction followed by *Mnl I* restriction enzyme digestion. *Results:* There was no difference in frequency of the 447Ter allele between patients (11.4%) and controls (11.0%). *Conclusions:* Either LPL does not play a role in AMD, or its effect is not revealed by this polymorphism or in this ethnic group.

Keywords: lipoprotein lipase, LPL, age related macular degeneration, AMD

---

* Correspondence to: Prof. Larry Baum, School of Pharmacy, Chinese University of Hong Kong, Shatin, Hong Kong, China, Tel. +852 39436833, Fax +852 26035295, E-mail lwbaum@cuhk.edu.hk

INTRODUCTION

Age-related macular degeneration (AMD) is a major cause of blindness in elderly, and occurs in 1% of Hong Kong Chinese residents aged over 40 [1]. The disease is accompanied by lipid-containing deposits called drusen beneath the retinal pigment epithelium (RPE), and atrophy of RPE. In the exudative form of AMD, the RPE detaches and choroidal neovascularization occurs [2].

Drusen contain vitronectin, apolipoprotein E (ApoE), complement proteins, ubiquitin, TIMP-3, lipids, cholesterol, carbohydrates, and many other components, sharing some similarities with atherosclerotic plaques [3-6]. Drusen may block the flow of oxygen, nutrients, and waste across Bruch's membrane (which separates the RPE from vessels in the choroid), thus damaging or killing RPE and photoreceptors [4].

Risk factors for AMD include age, smoking, and light skin and iris. Possible risk factors include hypertension, light exposure, body mass index, and blood fibrinogen levels [7-10].

The incidence of AMD is greater in relatives of patients, particularly in identical twins, suggesting a genetic factor [11;12]. There is evidence that variations in several genes may affect risk of AMD. One very large study revealed a significant association of ABCA4 gene alleles D2177N and G1961E with AMD [13]. The apparent function of ABCA4 as a flippase to remove N-retinylidene phosphatidylethanolamine from the lumenal to the cytoplasmic side of the rod disc membrane may be impaired by these mutations, resulting in light exposure increasing the concentration and protonation of this molecule, which is converted in retinal pigment epithelial cells (RPE) to N-retinylidene-N-retinylethanolamine (A2E). A2E accumulates in RPE and is cytotoxic, perhaps contributing to the RPE degeneration seen early in AMD [14-17]. There have been reports linking exudative AMD with polymorphisms in the paraoxonase and manganese superoxide dismutase genes [18, 19]. Both may function to protect cells from oxidative damage, which occurs in AMD and might be a causative factor [20]. Polymorphisms in paraoxonase, which is entirely bound to high density lipoprotein particles, have also been associated with risk of cardiovascular disease, perhaps through their ability to remove oxidized lipids from lipoprotein particles [17]. Another lipoprotein, ApoE, has also been genetically associated with AMD [21, 22]. ApoE transports lipids and cholesterol in the nervous system, and is highly expressed in response to neuronal damage [23;24]. The three major isoforms of ApoE: apo E2, E3 and, E4, are encoded by alleles ε2, ε3, and ε4 respectively. ApoE genotype is associated with dyslipidaemia, cardiovascular diseases and Alzheimer's disease [25]. Individuals with ε4 have increased risk for Alzheimer's disease but may have reduced risk of AMD [21, 22;26-28]. In addition, ApoE has been localized immunochemically to drusen [6, 22].

Lipoprotein lipase (LPL) promotes the binding of apoE to a major apoE receptor, LRP, and transfers apoE lipoprotein particle components to cells [29;30]. Since apoE polymorphisms may affect AMD risk, it is reasonable to ask whether LPL variants also affect AMD risk.

A polymorphism approximately 20 kb downstream from LPL, rs12678919, was associated with AMD in a GWAS, with an odds ratio of 1.26 [31]. Several other studies have since attempted to replicate this association, with mixed results [32-36].

One of the best characterized LPL polymorphisms is Ser447Ter. The 447Ter variant is missing two residues at the C-terminus, altering LPL activity, and resulting in lower blood levels of triglycerides and higher levels of HDL cholesterol [37-40]. This lipid profile is associated with decreased risk of heart disease, and people with the 447Ter allele have less coronary artery disease [41,42]. Depending on the ethnic group, 447Ter comprises 5-12% of alleles in the general population [38-40, 43;44]. We sought to determine whether the 447Ter is associated with AMD by genotyping a group of 140 Hong Kong Chinese AMD patients, of whom 17 had the exudative form of the disease, and 100 control subjects at least 70 years old.

METHODS

AMD patients and control subjects were recruited at the Prince of Wales Hospital Eye Clinic and Hong Kong Eye Hospital. All subjects gave informed consent. The study protocol was approved by the Ethics Committee on Human Research of the Chinese University of Hong Kong. All study subjects were given a complete ophthalmologic examination. Patients were diagnosed as having AMD by the presence of drusen and changes in the retinal pigment epithelium according to the findings of the fundus examination and if necessary with the help of fluorescein angiography. AMD with disciform scarring or neovascularization was classified as exudative (17 subjects). The controls (100 subjects) were free of eye disease except cataract and were at least 70 years of age.

Venous blood was drawn and DNA was extracted using a Qiagen blood kit. The Ser447Ter genotype was determined as previously described [45;46]. LPL exon 9 was amplified by polymerase chain reaction (PCR) using the primers TATTCACATCCATTTTCTTC and GTCAGCTTTAGCCCAGAATG, followed by *Mnl I* restriction enzyme digestion. A 3% agarose gel resolved fragments of 117 and 38 base pairs produced by mutant alleles. PCR reaction components were 2 μM each primer, 0.1 mM dNTP, 1.5 mM $MgCl_2$, 0.04 U Taq polymerase, and 1X Taq buffer. PCR was conducted for 30 cycles, at an annealing temperature of 50°C.

Differences between genotype and allele distributions in disease and controls were compared by chi-square or Fisher exact test.

RESULTS AND DISCUSSION

LPL Ser447Ter genotypes are listed in Table 1. Genotype and allele distributions were essentially the same for AMD vs. controls (p=0.94 for genotypes and p=0.88 for alleles), as well as for exudative AMD (p=0.89 for genotypes and p=1.00 for alleles).

The lack of an association of LPL 447Ter with AMD is consistent with a pathogenic mechanism of AMD which is not critically dependent on LPL activity, although a role for LPL cannot be entirely ruled out based on this negative result alone. For example, if LPL affects the development of only the exudative form of AMD, then our small sample of exudative cases may not provide sufficient power to detect the association.

Another possibility is that ethnic differences, either genetic or environmental, make variants of both ApoE and LPL risk factors in Europeans but not in Chinese. Supporting this suggestion is the fact that three studies in Europeans or European-Americans detected a decreased proportion of ApoE ε4 in AMD while the only study in Chinese failed to detect an association of ε4 with AMD [21;22;27;47]. Perhaps a higher intake of dietary fat in Europeans and Americans than in Chinese potentiates the effect of these lipoprotein-related gene polymorphisms on AMD risk. An association study of LPL 447Ter and AMD in Europeans may help to shed light on this possibility.

Other LPL alterations that affect enzymatic activity were not tested in this study. LPL with the 291Ser substitution exhibits 30-50% less activity than the wild-type enzyme, and people with this allele have higher blood triglyceride levels and lower HDL cholesterol levels [42; 48-52]. The 291Ser allele occurs in about 1-3.5% of the population in various ethnic groups [38, 48, 51-53]. However, it has not been reported in Chinese populations; therefore we did not genotype this polymorphism in the present study. This and other variants of LPL may nevertheless be worth investigating in relation to AMD in other ethnic groups.

Table 1. LPL 447 genotype and allele frequencies in AMD, exudative AMD, and normal controls

	AMD N (%)	Exudative AMD N (%)	Normal Controls N (%)
Genotypes			
Ser/Ser	109 (77.8)	14 (82.4) 82.4	79 (79.0) 79.0
Ser/Ter	30 (21.5) 21.5	3 (17.6)17.6	20 (20.0)
Ter/Ter	1 (0.7) 0.7	0 (0.0) 0.0	1 (1.0)
Total	140	17	100
Alleles			
Ser	248 (88.6) 88.6	31 (91.2) 91.2	178 (89.0)
Ter	32 (11.4) 11.4	3 (8.8) 8.8	22 (11.0) 11.0
Total	280	34	200

References

[1] Van Newkirk, MR. The Hong Kong vision study: A pilot assessment of visual impairment in adults. *Trans Am. Ophthalmol. Soc.,* 1997 95, 715-749.

[2] Kanski, JJ. *Clinical Ophthalmology.* 3rd, revised. Oxford: Butterworth-Heinemann; 1997.

[3] Mullins, RF; Russell, SR; Anderson, DH; Hageman, GS. Drusen associated with aging and age-related macular degeneration contain proteins common to extracellular deposits associated with atherosclerosis, elastosis, amyloidosis, and dense deposit disease. *FASEB J.,* 2000 14, 835-846.

[4] Hageman, GS; Mullins, RF. Molecular composition of drusen as related to substructural phenotype. *Mol. Vis.,* 1999 5, 28.

[5] Curcio, CA; Millican, CL; Bailey, T; Kruth, HS. Accumulation of cholesterol with age in human bruch's membrane. *Invest. Ophthalmol. Vis. Sci.,* 2001 42, 265-274.

[6] Curcio, CA; Johnson, M; Rudolf, M; Huang, JD. The oil spill in ageing bruch membrane. *Br. J. Ophthalmol.,* 2011 95, 1638-1645.

[7] Yates, JR; Moore, AT. Genetic susceptibility to age related macular degeneration. *J. Med. Genet.*, 2000 37, 83-87.
[8] Smith, W; Mitchell, P; Leeder, SR; Wang, JJ. Plasma fibrinogen levels, other cardiovascular risk factors, and age-related maculopathy: The Blue Mountains Eye Study. *Arch. Ophthalmol.*, 1998 116, 583-587.
[9] Hawkins, BS; Bird, A; Klein, R; West, SK. Epidemiology of age-related macular degeneration. *Mol. Vis.*, 1999 5, 26.
[10] Age-Related Eye Disease Study Research Group. Risk factors associated with age-related macular degeneration. A case-control study in the age-related eye disease study: Age-related eye disease study report number 3. *Ophthalmology*, 2000 107, 2224-2232.
[11] Gorin, MB; Breitner, JC; De Jong, PT; Hageman, GS; Klaver, CC; Kuehn, MH; Seddon, JM. The genetics of age-related macular degeneration. *Mol. Vis.*, 1999 5, 29.
[12] Seddon, JM; Ajani, UA; Mitchell, BD. Familial aggregation of age-related maculopathy. *Am. J. Ophthalmol.*, 1997 123, 199-206.
[13] Allikmets, R. Further evidence for an association of ABCR alleles with age-related macular degeneration. the international ABCR screening consortium. *Am. J. Hum. Genet.*, 2000 67, 487-491.
[14] Mata, NL; Weng, J; Travis, GH. Biosynthesis of a major lipofuscin fluorophore in mice and humans with ABCR-mediated retinal and macular degeneration. *Proc. Natl. Acad. Sci. U S A*, 2000 97, 7154-7159.
[15] Weng, J; Mata, NL; Azarian, SM; Tzekov, RT; Birch, DG; Travis, GH. Insights into the function of rim protein in photoreceptors and etiology of stargardt's disease from the phenotype in abcr knockout mice. *Cell*, 1999 98, 13-23.
[16] Ahn, J; Wong, JT; Molday, RS. The effect of lipid environment and retinoids on the ATPase activity of ABCR, the photoreceptor ABC transporter responsible for stargardt macular dystrophy. *J. Biol. Chem.*, 2000 275, 20399-20405.
[17] Garin, MC; James, RW; Dussoix, P; Blanche, H; Passa, P; Froguel, P; Ruiz, J. Paraoxonase polymorphism met-Leu54 is associated with modified serum concentrations of the enzyme. A possible link between the paraoxonase gene and increased risk of cardiovascular disease in diabetes. *J. Clin. Invest.*, 1997 99, 62-66.
[18] Ikeda, T; Obayashi, H; Hasegawa, G; Nakamura, N; Yoshikawa, T; Imamura, Y; Koizumi, K; Kinoshita, S. Paraoxonase gene polymorphisms and plasma oxidized low-density lipoprotein level as possible risk factors for exudative age-related macular degeneration. *Am. J. Ophthalmol.*, 2001 132, 191-195.
[19] Kimura, K; Isashiki, Y; Sonoda, S; Kakiuchi-Matsumoto, T; Ohba, N. Genetic association of manganese superoxide dismutase with exudative age-related macular degeneration. *Am. J. Ophthalmol.*, 2000 130, 769-773.
[20] Winkler, BS; Boulton, ME; Gottsch, JD; Sternberg, P. Oxidative damage and age-related macular degeneration. *Mol. Vis.*, 1999 5, 32.
[21] Souied, EH; Benlian, P; Amouyel, P; Feingold, J; Lagarde, JP; Munnich, A; Kaplan, J; Coscas, G; Soubrane, G. The epsilon4 allele of the apolipoprotein E gene as a potential protective factor for exudative age-related macular degeneration. *Am. J. Ophthalmol.*, 1998 125, 353-359.
[22] Klaver, CC; Kliffen, M; van Duijn, CM; Hofman, A; Cruts, M; Grobbee, DE; van Broeckhoven, C; de Jong, PT. Genetic association of apolipoprotein E with age-related macular degeneration. *Am. J. Hum. Genet.*, 1998 63, 200-206.

[23] Ignatius, MJ; Gebicke-Harter, PJ; Skene, JH; Schilling, JW; Weisgraber, KH; Mahley, RW; Shooter, EM. Expression of apolipoprotein E during nerve degeneration and regeneration. *Proc. Natl. Acad. Sci. U S A,* 1986 83, 1125-1129.

[24] Handelmann, GE; Boyles, JK; Weisgraber, KH; Mahley, RW; Pitas, RE. Effects of apolipoprotein E, beta-very low density lipoproteins, and cholesterol on the extension of neurites by rabbit dorsal root ganglion neurons in vitro. *J. Lipid. Res.,* 1992 33, 1677-1688.

[25] Siest, G; Pillot, T; Regis-Bailly, A; Leininger-Muller, B; Steinmetz, J; Galteau, MM; Visvikis, S. Apolipoprotein E: An important gene and protein to follow in laboratory medicine. *Clin. Chem.,* 1995 41, 1068-1086.

[26] Corder, EH; Saunders, AM; Strittmatter, WJ; Schmechel, DE; Gaskell, PC; Small, GW; Roses, AD; Haines, JL; Pericak-Vance, MA. Gene dose of apolipoprotein E type 4 allele and the risk of Alzheimer's disease in late onset families. *Science,* 1993 261, 921-923.

[27] Simonelli, F; Margaglione, M; Testa, F; Cappucci, G; Manitto, MP; Brancato, R; Rinaldi, E. Apolipoprotein E polymorphisms in age-related macular degeneration in an italian population. *Ophthalmic. Res.,* 2001 33, 325-328.

[28] Thakkinstian, A; Bowe, S; McEvoy, M; Smith, W; Attia, J. Association between apolipoprotein E polymorphisms and age-related macular degeneration: A HuGE review and meta-analysis. *Am. J. Epidemiol.,* 2006 164, 813-822.

[29] Rebeck, GW; Harr, SD; Strickland, DK; Hyman, BT. Multiple, diverse senile plaque-associated proteins are ligands of an apolipoprotein E receptor, the alpha 2-macroglobulin receptor/low-density-lipoprotein receptor-related protein. *Ann. Neurol.,* 1995 37, 211-217.

[30] Strickland, DK; Kounnas, MZ; Argraves, WS. LDL receptor-related protein: A multiligand receptor for lipoprotein and proteinase catabolism. *FASEB J.,* 1995 9, 890-898.

[31] Chen, W; Stambolian, D; Edwards, AO; Branham, KE; Othman, M; Jakobsdottir, J; Tosakulwong, N; Pericak-Vance, MA; Campochiaro, PA; Klein, ML; Tan, PL; Conley, YP; Kanda, A; Kopplin, L; Li, Y; Augustaitis, KJ; Karoukis, AJ; Scott, WK; Agarwal, A; Kovach, JL; Schwartz, SG; Postel, EA; Brooks, M; Baratz, KH; Brown, WL; Complications of Age-Related Macular Degeneration Prevention Trial Research Group; Brucker, AJ; Orlin, A; Brown, G; Ho, A; Regillo, C; Donoso, L; Tian, L; Kaderli, B; Hadley, D; Hagstrom, SA; Peachey, NS; Klein, R; Klein, BE; Gotoh, N; Yamashiro, K; Ferris Iii, F; Fagerness, JA; Reynolds, R; Farrer, LA; Kim, IK; Miller, JW; Corton, M; Carracedo, A; Sanchez-Salorio, M; Pugh, EW; Doheny, KF; Brion, M; Deangelis, MM; Weeks, DE; Zack, DJ; Chew, EY; Heckenlively, JR; Yoshimura, N; Iyengar, SK; Francis, PJ; Katsanis, N; Seddon, JM; Haines, JL; Gorin, MB; Abecasis, GR; Swaroop, A. Genetic variants near TIMP3 and high-density lipoprotein-associated loci influence susceptibility to age-related macular degeneration. *Proc. Natl. Acad. Sci. USA,* 2010 107, 7401-7406.

[32] Fauser, S; Smailhodzic, D; Caramoy, A; van de Ven, JP; Kirchhof, B; Hoyng, CB; Klevering, BJ; Liakopoulos, S; den Hollander, AI. Evaluation of serum lipid concentrations and genetic variants at high-density lipoprotein metabolism loci and TIMP3 in age-related macular degeneration. *Invest. Ophthalmol. Vis. Sci.,* 2011 52, 5525-5528.

[33] Peter, I; Huggins, GS; Ordovas, JM; Haan, M; Seddon, JM. Evaluation of new and established age-related macular degeneration susceptibility genes in the women's health initiative sight exam (WHI-SE) study. *Am. J. Ophthalmol.,* 2011 152, 1005-1013.e1.

[34] Tian, J; Yu, W; Qin, X; Fang, K; Chen, Q; Hou, J; Li, J; Chen, D; Hu, Y; Li, X. Association of genetic polymorphisms and age-related macular degeneration in chinese population. *Invest. Ophthalmol. Vis. Sci.,* 2012 53, 4262-4269.

[35] Zhang, X; Li, M; Wen, F; Zuo, C; Chen, H; Wu, K; Zeng, R. Different impact of high-density lipoprotein-related genetic variants on polypoidal choroidal vasculopathy and neovascular age-related macular degeneration in a chinese han population. *Exp. Eye Res.,* 2013 108, 16-22.

[36] Lechanteur, YT; van de Ven, JP; Smailhodzic, D; Boon, CJ; Klevering, BJ; Fauser, S; Groenewoud, JM; van der Wilt, GJ; den Hollander, AI; Hoyng, CB. Genetic, behavioral, and sociodemographic risk factors for second eye progression in age-related macular degeneration. *Invest. Ophthalmol. Vis. Sci.,* 2012 53, 5846-5852.

[37] Groenemeijer, BE; Hallman, MD; Reymer, PW; Gagne, E; Kuivenhoven, JA; Bruin, T; Jansen, H; Lie, KI; Bruschke, AV; Boerwinkle, E; Hayden, MR; Kastelein, JJ. Genetic variant showing a positive interaction with beta-blocking agents with a beneficial influence on lipoprotein lipase activity, HDL cholesterol, and triglyceride levels in coronary artery disease patients. The Ser447-stop substitution in the lipoprotein lipase gene. REGRESS study group. *Circulation,* 1997 95, 2628-2635.

[38] Jemaa, R; Fumeron, F; Poirier, O; Lecerf, L; Evans, A; Arveiler, D; Luc, G; Cambou, JP; Bard, JM; Fruchart, JC; Apfelbaum, M; Cambien, F; Tiret, L. Lipoprotein lipase gene polymorphisms: Associations with myocardial infarction and lipoprotein levels, the ECTIM study. Etude Cas Témoin sur l'infarctus du Myocarde. *J. Lipid. Res.,* 1995 36, 2141-2146.

[39] Stocks, J; Thorn, JA; Galton, DJ. Lipoprotein lipase genotypes for a common premature termination codon mutation detected by PCR-mediated site-directed mutagenesis and restriction digestion. *J. Lipid. Res.,* 1992 33, 853-857.

[40] Zhang, Q; Cavanna, J; Winkelman, BR; Shine, B; Gross, W; Marz, W; Galton, DJ. Common genetic variants of lipoprotein lipase that relate to lipid transport in patients with premature coronary artery disease. *Clin. Genet.,* 1995 48, 293-298.

[41] Galton, DJ; Mattu, R; Needham, EW; Cavanna, J. Identification of putative beneficial mutations for lipid transport. *Z Gastroenterol,* 1996 34 Suppl 3, 56-58.

[42] Wittrup, HH; Tybjaerg-Hansen, A; Nordestgaard, BG. Lipoprotein lipase mutations, plasma lipids and lipoproteins, and risk of ischemic heart disease. A meta-analysis. *Circulation,* 1999 99, 2901-2907.

[43] Peacock, RE; Hamsten, A; Nilsson-Ehle, P; Humphries, SE. Associations between lipoprotein lipase gene polymorphisms and plasma correlations of lipids, lipoproteins and lipase activities in young myocardial infarction survivors and age-matched healthy individuals from sweden. *Atherosclerosis,* 1992 97, 171-185.

[44] Wiebusch, H; Funke, H; Saku, K; Arakawa, K; Assmann, G. Mutational analysis of two frequently observed lipoprotein lipase gene variants in the japanese population. *J. Inherit. Metab. Dis.,* 1997 20, 108-109.

[45] Baum, L; Chen, L; Masliah, E; Chan, YS; Ng, HK; Pang, CP. Lipoprotein lipase mutations and Alzheimer's disease. *Am. J. Med. Genet.,* 1999 88, 136-139.

[46] Baum, L; Wiebusch, H; Pang, CP. Roles for lipoprotein lipase in Alzheimer's disease: An association study. *Microsc Res Tech,* 2000 50, 291-296.

[47] Pang, CP; Baum, L; Chan, WM; Lau, TC; Poon, PM; Lam, DS. The apolipoprotein E epsilon4 allele is unlikely to be a major risk factor of age-related macular degeneration in Chinese. *Ophthalmologica,* 2000 214, 289-291.

[48] Fisher, RM; Mailly, F; Peacock, RE; Hamsten, A; Seed, M; Yudkin, JS; Beisiegel, U; Feussner, G; Miller, G; Humphries, SE. Interaction of the lipoprotein lipase asparagine 291-->serine mutation with body mass index determines elevated plasma triacylglycerol concentrations: A study in hyperlipidemic subjects, myocardial infarction survivors, and healthy adults. *J. Lipid. Res.,* 1995 36, 2104-2112.

[49] Hoffer, MJ; Bredie, SJ; Boomsma, DI; Reymer, PW; Kastelein, JJ; de Knijff, P; Demacker, PN; Stalenhoef, AF; Havekes, LM; Frants, RR. The lipoprotein lipase (Asn291-->Ser) mutation is associated with elevated lipid levels in families with familial combined hyperlipidaemia. *Atherosclerosis,* 1996 119, 159-167.

[50] Pimstone, SN; Clee, SM; Gagne, SE; Miao, L; Zhang, H; Stein, EA; Hayden, MR. A frequently occurring mutation in the lipoprotein lipase gene (Asn291Ser) results in altered postprandial chylomicron triglyceride and retinyl palmitate response in normolipidemic carriers. *J. Lipid. Res.,* 1996 37, 1675-1684.

[51] Reymer, PW; Gagne, E; Groenemeyer, BE; Zhang, H; Forsyth, I; Jansen, H; Seidell, JC; Kromhout, D; Lie, KE; Kastelein, J; Hayden, MR. A lipoprotein lipase mutation (Asn291Ser) is associated with reduced HDL cholesterol levels in premature atherosclerosis. *Nat. Genet.,* 1995 10, 28-34.

[52] Reymer, PW; Groenemeyer, BE; Gagne, E; Miao, L; Appelman, EE; Seidel, JC; Kromhout, D; Bijvoet, SM; van de Oever, K; Bruin, T. A frequently occurring mutation in the lipoprotein lipase gene (Asn291Ser) contributes to the expression of familial combined hyperlipidemia. *Hum. Mol. Genet.,* 1995 4, 1543-1549.

[53] Zhang, H; Reymer, PW; Liu, MS; Forsythe, IJ; Groenemeyer, BE; Frohlich, J; Brunzell, JD; Kastelein, JJ; Hayden, MR; Ma, Y. Patients with apoE3 deficiency (E2/2, E3/2, and E4/2) who manifest with hyperlipidemia have increased frequency of an Asn 291-->Ser mutation in the human LPL gene. *Arterioscler. Thromb. Vasc. Biol.,* 1995 15, 1695-1703.

In: Biochemistry and Molecular Biology
Editor: Mary Miu Yee Waye

ISBN: 978-1-63482-312-8

Chaper 4

A CASE-CONTROL STUDY OF ASSOCIATION BETWEEN BRAIN-DERIVED NEUROTROPHIC FACTOR 712G>A POLYMORPHISM AND PARKINSON'S DISEASE

***Qiwen Liao*[1], *Peter Y. K. Poon*[2], *Anne Y. Y. Chan*[2], *Vincent C. T. Mok*[2] *and Larry Baum*[3*]**

[1]Division of Genomics and Bioinformatics,
CUHK-BGI Innovation Institute of Trans-Omics,
The Chinese University of Hong Kong, Shatin, Hong Kong
[2]Division of Neurology, Department of Medicine and Therapeutics,
The Chinese University of Hong Kong, Shatin, Hong Kong
[3]School of Pharmacy, The Chinese University of Hong Kong, Shatin, Hong Kong

ABSTRACT

Parkinson's disease (PD) is a progressive neurodegenerative disorder. A study reported a strong association of PD with the A allele of the -712G>A polymorphism in the Brain-derived neurotrophic factor (BDNF) gene. Therefore, we sought to confirm this finding. 261 PD patients and 199 healthy controls were genotyped using amplification refractory mutation system polymerase chain reaction (ARMS PCR). Statistical analysis was performed using the Epitools package in R software version 3.0.1. The A allele of BDNF -712G>A did not display a significant association with PD (OR = 1.00, 95% CI = 0.52-1.95, p=1.0). The BDNF -712G>A polymorphism may not affect the risk of developing PD. Further studies may help resolve the discrepancy between the previous and current reports of this association.

Keywords: Parkinson's disease; Brain-derived neurotrophic factor; polymorphism

* Correspondence to: Prof. Larry Baum, School of Pharmacy. The Chinese University of Hong Kong. Shatin, Hong Kong, China. Phone: 852-39436833. Fax: 852-26035295. E-mail: lwbaum@cuhk.edu.hk

INTRODUCTION

Parkinson's disease (PD) is a common progressive neurodegenerative disorder [1]. It is characterized clinically by Parkinsonism (tremor, rigidity, bradykinesia and postural instability) and pathologically by the loss of dopaminergic neurons in the substantia nigra and by the presence of intraneuronal deposits of α-synuclein protein called Lewy bodies [2-4]. As with other neurodegenerative diseases, aging is the major risk factor, with a median age of onset of 60 years [1, 5]. Genes and environmental factors can interact to modulate the risk of disease [6-8]. Genetic studies of PD have attempted to identify causative genes in PD families as well as susceptibility genes in sporadic cases. These genes include SNCA, LRRK2, PRKN, PINK1, DJ-1, and GBA [3, 9].

Brain-derived neurotrophic factor (BDNF) is a member of a family of structurally related neurotrophic factors. It is the most abundant neurotrophin in the brain and promotes neuronal survival, differentiation and synaptic plasticity [10-13]. Therefore, BDNF might play a role in neurodegenerative diseases (such as Huntington disease and Alzheimer's disease) and psychiatric disorders (such as major depressive disorder and schizophrenia) [13-15].

BDNF mRNA is strongly expressed by dopaminergic neurons in the substantia nigra pars compacta, as revealed by situ hybridization [16]. BDNF enhances the survival of dopaminergic neurons and protects them against the neurotoxic effects of 1-methyl-4-phenylpyridinium (MPP+) and 6-hydroxydopamine (6-OHDA) [17, 18]. In nigrostriatal regions, the concentration of BDNF protein, as measured by enzyme-linked immunosorbent assay (ELISA), was significantly decreased in PD [19]. Studies of the association of PD with BDNF genetic variants, principally -196G>A and 270C>T, were inconsistent [11, 20-26]. However, one study of BDNF polymorphisms reported a very strong association of PD with -712G>A [27], thus we sought to confirm this finding in a case-control study.

MATERIALS AND METHODS

We included 261 PD patients [99 females and 162 males, aged 63.4 ± 8.0 years (mean ± SD)] and 199 controls [70 females and 129 males, aged 62.2 ± 6.8 years (mean ± SD)], mostly from a previous study [28]. All subjects were unrelated to any other subjects and were Hong Kong Chinese. Controls had no evidence of Parkinsonism or cognitive impairment on clinical examination, and were recruited from the hospital staff and community. The diagnosis of PD was made in accordance with United Kingdom Parkinson's Disease Society Brain Bank criteria [29]. All participants were examined by movement disorder specialists (A.Y.Y.C. and V.C.T.M.). Informed consent was obtained from each participant.

DNA was purified from blood samples. Genotyping was performed by the tetra-primer amplification refractory mutation system polymerase chain reaction (ARMS PCR) method [30-33]. Four primers were used to produce PCR products of different lengths for each of the two alleles. Primers were designed by the BatchPrimer3 software [32] and are listed in Table 1. PCR reactions (total volume 20 μl) contained 50 ng of genomic DNA, 2.5 mM $MgCl_2$, 200 μM dNTP, and 1 U of Taq DNA polymerase (Roche) with PCR buffer and the four primers at the concentrations listed in Table 1. Cycling conditions were 2 min at 95°C, followed by 35 cycles of 95°C for 1 minute, annealing at 64°C for 1 minute and 72°C for 1 minute, and a final cycle at 72°C for 2 minutes. PCR products were separated by electrophoresis on 3% agarose gels. We set the level of statistical significance as $p<0.05$. Hardy-Weinberg equilibrium of the genotype distribution of controls was evaluated by a chi-square test using the SHEsis program [34]. The epitools package of R software version 3.0.1 was used to calculate the frequencies of alleles and genotypes, odds ratios (OR), and 95% confidence intervals (95% CI). SPSS version 18 software was used to perform t-tests.

Table 1. Genotyping primers

Primer	Sequence	Concentration (nM)
Forward inner (A allele)	GGGGGCGTTCATCAATAAAAAATGATA	75
Reverse inner (G allele)	GCCTAATAGGAATTCAATCAAGATAATCAC	625
Forward outer	GTTTTTCTGTTTGGTCGGCTAGAAAG	62
Reverse outer	GGCTATTGGTTTACTCTGCTGCCTTA	62

RESULTS

The allele and genotype distributions of the BDNF -712G>A polymorphism in PD patients and controls are summarized in Table 2. The genotype distribution in controls was examined for consistency with the Hardy-Weinberg equilibrium (HWE), and no significant deviation from HWE was detected: p=0.21.

Table 2. The distribution of BDNF-712G>A genotypes and alleles in our case-control study

BDNF-712G>A	PD[a]	Control	OR (95% CI)[b]	P-value
GG	241 (0.923)	184 (0.925)	1.00 (ref)	0.96
GA	19 (0.073)	14 (0.070)		
AA	1 (0.004)	1 (0.005)		
GA/AA	20 (0.077)	15 (0.075)	1.02 (0.51-2.04)	
G	501 (0.960)	382 (0.960)	1.00 (ref)	1.00
A	21 (0.040)	16 (0.040)	1.00 (0.52-1.95)	

[a] PD: Parkinson's disease. [b] OR (95% CI): odds ratio (95% confidence interval)

Neither A+ genotypes (the combination of GA and AA) nor A alleles were significantly associated with Parkinson's disease. Stratification by gender in Table 3 revealed no significant associations of genotypes with PD, though there were opposite tendencies in the genders: toward association of A+ genotypes with PD in males and association of the GG genotype with PD in females.

Table 3. Gender distribution of BDNF-712G>A genotypes

BDNF-712G>A	PD[a]	Control	OR (95%CI)[b]	P-value
Females				
GG	93 (0.939)	62 (0.886)	1.00 (ref)	0.26
GA/AA	6 (0.061)	8 (0.114)	0.50 (0.17-1.51)	
Males				
GG	148 (0.914)	122 (0.946)	1.00 (ref)	0.36
GA/AA	14 (0.086)	7 (0.054)	1.65 (0.65-4.21)	

[a] PD: Parkinson's disease. [b] OR (95% CI): odds ratio (95% confidence interval)

Onset age of PD was known for 249 of the patients. Mean onset age was 56.9 years (standard deviation: 10.5 years) for patients with the GG genotype and 55.2 years (standard deviation: 11.1 years) for patients with A+ genotypes. The difference in onset ages between GG and A+ was not significant: p=0.52.

Discussion

PD is the second most common age-related neurodegenerative disorder [35]. Analyses of brain tissue from patients with PD implicate the death of dopamine neurons [36;37]. Decreased brain levels of BDNF were detected in postmortem studies of PD patients [38]. The BDNF -712G>A polymorphism is located in the BDNF promoter region, 712 bases upstream of the first exon. This region is potentially a part of the eukaryotic polymerase II promoter binding site, and the A allele may disrupt the pattern of recognition and thus reduce BDNF expression [39;40]. Thus, studies on the association between the BDNF -712G>A polymorphism and PD may help determine whether BDNF plays a role in PD.

One paper has reported an association of BDNF -712G>A with PD, revealing a very strong effect: an odds ratio of 3.2 ($p<0.001$) for the A allele [27]. However, no further reports have examined this association. Our study found no significant relation between genotypes and either age of onset or gender, and no evidence of association of the A allele with PD, at least within the bounds of the 95% confidence interval: 0.52-1.95. Our result is thus inconsistent with the previous finding.

There are several possible explanations for this discrepancy. Experimental error might have occurred in either study. We used ARMS PCR, while the above study used PCR RFLP. In this PCR RFLP assay, for example, incomplete digestion by the restriction endonuclease in a subset of samples, such as patients, would have resulted in an artifactual excess of the AA genotype in that subset. Hardy-Weinberg equilibrium testing of patients in the PCR RFLP study reveals a p value of 0.014 and double the expected number of patients with the AA genotype (9 vs. 4.5), while no deviation from Hardy-Weinberg equilibrium was seen in controls (p=0.3) or in our study (p=0.4 in patients and p=0.2 in controls). Because AA is the rarest genotype and thus would be the cheapest genotype to sequence, and because the difference between patients and controls in the above paper was greater for the AA genotype than between other genotypes, sequencing samples with the AA genotype may help reveal such error and thus resolve the discrepancy between the two reports.

Another possibility is that cryptic sub-ethnic stratification took place. Subjects for the first study were Chinese from Nanjing, while our subjects were Chinese from Hong Kong. If genotypes of this polymorphism are present at different frequencies among people from different regions in China, then a preponderance of subjects from one region among patients compared to controls may create an apparent association with PD. Further studies in other regions or ethnic groups may reduce this possible artifact.

A third and related explanation could be linkage disequilibrium that varies among regions in China. If the -712G>A polymorphism is a tagging polymorphism in linkage disequilibrium with the causative variant, and if the linkage disequilibrium is strong in Nanjing but not Hong Kong, then the association would be detectable in Nanjing but not in Hong Kong. As with the above possibility, further studies in other regions may explore this.

Our findings suggest that, at least within the ethnic scope of our study, this polymorphism of BDNF might not alter the risk of developing PD. If BDNF is found not to affect PD via genetic variations, investigation may still be warranted into the etiologic and potential therapeutic role of this important neurotrophic factor in PD. Supplementing BDNF supply to the substantia nigra and striatum might stave off dopaminergic loss and help treat the disease [41].

REFERENCES

[1] Lees, AJ; Hardy, J; Revesz, T. Parkinson's disease. *Lancet,* 2009 373, 2055-2066.

[2] Nussbaum, RL; Ellis, CE. Alzheimer's disease and Parkinson's disease. *N Engl J Med,* 2003 348, 1356-1364.

[3] Chung, SJ; Jung, Y; Hong, M; Kim, MJ; You, S; Kim, YJ; Kim, J; Song, K. Alzheimer's disease and parkinson's disease genome-wide association study top hits and risk of parkinson's disease in korean population. *Neurobiol Aging,* 2013 34, 2695.e1-2695.e7.

[4] Udupa, K; Chen, R. Motor cortical plasticity in parkinson's disease. *Front Neurol,* 2013 4, 128.

[5] de Lau, LM; Breteler, MM. Epidemiology of parkinson's disease. *Lancet Neurol,* 2006 5, 525-535.

[6] Dai, L; Wang, D; Meng, H; Zhang, K; Fu, L; Wu, Y; Bai, Y. Association between the BDNF G196A and C270T polymorphisms and parkinson's disease: A meta-analysis. *Int J Neurosci,* 2013 123, 675-683.

[7] Peeraully, T; Tan, EK. Genetic variants in sporadic parkinson's disease: East vs west. *Parkinsonism Relat Disord,* 2012 18 Suppl 1, S63-5.

[8] Tian, YY; Tang, CJ; Wu, J; Zhou, JS. Parkinson's disease in china. *Neurol Sci,* 2011 32, 23-30.

[9] Houlden, H; Singleton, AB. The genetics and neuropathology of Parkinson's disease. *Acta Neuropathol,* 2012 124, 325-338.

[10] Gall, CM; Gold, SJ; Isackson, PJ; Seroogy, KB. Brain-derived neurotrophic factor and neurotrophin-3 mRNAs are expressed in ventral midbrain regions containing dopaminergic neurons. *Mol Cell Neurosci,* 1992 3, 56-63.

[11] Hong, CJ; Liou, YJ; Tsai, SJ. Effects of BDNF polymorphisms on brain function and behavior in health and disease. *Brain Res Bull,* 2011 86, 287-297.
[12] Luhder, F; Gold, R; Flugel, A; Linker, RA. Brain-derived neurotrophic factor in neuroimmunology: Lessons learned from multiple sclerosis patients and experimental autoimmune encephalomyelitis models. *Arch Immunol Ther Exp (Warsz),* 2013 61, 95-105.
[13] Zuccato, C; Cattaneo, E. Brain-derived neurotrophic factor in neurodegenerative diseases. *Nat Rev Neurol,* 2009 5, 311-322.
[14] Nagahara, AH; Tuszynski, MH. Potential therapeutic uses of BDNF in neurological and psychiatric disorders. *Nat Rev Drug Discov,* 2011 10, 209-219.
[15] Weickert, CS; Hyde, TM; Lipska, BK; Herman, MM; Weinberger, DR; Kleinman, JE. Reduced brain-derived neurotrophic factor in prefrontal cortex of patients with schizophrenia. *Mol Psychiatry,* 2003 8, 592-610.
[16] Howells, DW; Porritt, MJ; Wong, JY; Batchelor, PE; Kalnins, R; Hughes, AJ; Donnan, GA. Reduced BDNF mRNA expression in the parkinson's disease substantia nigra. *Exp Neurol,* 2000 166, 127-135.
[17] Parsian, A; Sinha, R; Racette, B; Zhao, JH; Perlmutter, JS. Association of a variation in the promoter region of the brain-derived neurotrophic factor gene with familial Parkinson's disease. *Parkinsonism Relat Disord,* 2004 10, 213-219.
[18] Volpe, BT; Wildmann, J; Altar, CA. Brain-derived neurotrophic factor prevents the loss of nigral neurons induced by excitotoxic striatal-pallidal lesions. *Neuroscience,* 1998 83, 741-748.
[19] Mogi, M; Togari, A; Kondo, T; Mizuno, Y; Komure, O; Kuno, S; Ichinose, H; Nagatsu, T. Brain-derived growth factor and nerve growth factor concentrations are decreased in the substantia nigra in parkinson's disease. *Neurosci Lett,* 1999 270, 45-48.
[20] Svetel, M; Pekmezovic, T; Markovic, V; Novakovic, I; Dobricic, V; Djuric, G; Stefanova, E; Kostic, V. No association between brain-derived neurotrophic factor G196A polymorphism and clinical features of Parkinson's disease. *Eur Neurol,* 2013 70, 257-262.
[21] Foltynie, T; Lewis, SG; Goldberg, TE; Blackwell, AD; Kolachana, BS; Weinberger, DR; Robbins, TW; Barker, RA. The BDNF Val66Met polymorphism has a gender specific influence on planning ability in Parkinson's disease. *J Neurol,* 2005 252, 833-838.
[22] Zintzaras, E; Hadjigeorgiou, GM. The role of G196A polymorphism in the brain-derived neurotrophic factor gene in the cause of Parkinson's disease: A meta-analysis. *J Hum Genet,* 2005 50, 560-566.
[23] Gao, L; Diaz-Corrales, FJ; Carrillo, F; Diaz-Martin, J; Caceres-Redondo, MT; Carballo, M; Palomino, A; Lopez-Barneo, J; Mir, P. Brain-derived neurotrophic factor G196A polymorphism and clinical features in parkinson's disease. *Acta Neurol Scand,* 2010 122, 41-45.
[24] Guerini, FR; Beghi, E; Riboldazzi, G; Zangaglia, R; Pianezzola, C; Bono, G; Casali, C; Di Lorenzo, C; Agliardi, C; Nappi, G; Clerici, M; Martignoni, E. BDNF Val66Met polymorphism is associated with cognitive impairment in italian patients with parkinson's disease. *Eur J Neurol,* 2009 16, 1240-1245.
[25] Saarela, MS; Lehtimaki, T; Rinne, JO; Huhtala, H; Rontu, R; Hervonen, A; Roytta, M; Ahonen, JP; Mattila, KM. No association between the brain-derived neurotrophic factor

196 G>A or 270 C>T polymorphisms and alzheimer's or Parkinson's disease. *Folia Neuropathol,* 2006 44, 12-16.
[26] Hong, CJ; Liu, HC; Liu, TY; Lin, CH; Cheng, CY; Tsai, SJ. Brain-derived neurotrophic factor (BDNF) Val66Met polymorphisms in Parkinson's disease and age of onset. *Neurosci Lett,* 2003 353, 75-77.
[27] Chen, L; Wang, Y; Xiao, H; Wang, L; Wang, C; Guo, S; Zhao, Y; Hua, P; Liu, W; Zhang, N. The 712A/G polymorphism of brain-derived neurotrophic factor is associated with parkinson's disease but not major depressive disorder in a chinese han population. *Biochem Biophys Res Commun,* 2011 408, 318-321.
[28] Chan, AY; Baum, L; Tang, NL; Lau, CY; Ng, PW; Hui, KF; Mizuno, Y; Kwan, JY; Mok, VC; Kuo, SH. The role of the Ala746Thr variant in the ATP13A2 gene among chinese patients with parkinson's disease. *J Clin Neurosci,* 2013 20, 761-762.
[29] Hughes, AJ; Daniel, SE; Kilford, L; Lees, AJ. Accuracy of clinical diagnosis of idiopathic Parkinson's disease: A clinico-pathological study of 100 cases. *J Neurol Neurosurg Psychiatry,* 1992 55, 181-184.
[30] Ye, S; Humphries, S; Green, F. Allele specific amplification by tetra-primer PCR. *Nucleic Acids Res,* 1992 20, 1152.
[31] Chiapparino, E; Lee, D; Donini, P. Genotyping single nucleotide polymorphisms in barley by tetra-primer ARMS-PCR. *Genome,* 2004 47, 414-420.
[32] Okayama, N; Fujimura, K; Nakamura, J; Suehiro, Y; Hamanaka, Y; Hinoda, Y. Evaluation of a new efficient procedure for single-nucleotide polymorphism genotyping: Tetra-primer amplification refractory mutation system-polymerase chain reaction. *Clin Chem Lab Med,* 2004 42, 13-16.
[33] You, FM; Huo, N; Gu, YQ; Luo, MC; Ma, Y; Hane, D; Lazo, GR; Dvorak, J; Anderson, OD. BatchPrimer3: A high throughput web application for PCR and sequencing primer design. *BMC Bioinformatics,* 2008 9, 253-2105-9-253.
[34] Shi, YY; He, L. SHEsis, a powerful software platform for analyses of linkage disequilibrium, haplotype construction, and genetic association at polymorphism loci. *Cell Res,* 2005 15, 97-98.
[35] Olanow, CW; Stern, MB; Sethi, K. The scientific and clinical basis for the treatment of Parkinson disease (2009). *Neurology,* 2009 72, S1-136.
[36] Sterling, NW; Du, G; Lewis, MM; Dimaio, C; Kong, L; Eslinger, PJ; Styner, M; Huang, X. Striatal shape in parkinson's disease. *Neurobiol Aging,* 2013 34, 2510-2516.
[37] Mattson, MP. Apoptosis in neurodegenerative disorders. *Nat Rev Mol Cell Biol,* 2000 1, 120-129.
[38] Baquet, ZC; Bickford, PC; Jones, KR. Brain-derived neurotrophic factor is required for the establishment of the proper number of dopaminergic neurons in the substantia nigra pars compacta. *J Neurosci,* 2005 25, 6251-6259.
[39] Zhang, H; Ozbay, F; Lappalainen, J; Kranzler, HR; van Dyck, CH; Charney, DS; Price, LH; Southwick, S; Yang, BZ; Rasmussen, A; Gelernter, J. Brain derived neurotrophic factor (BDNF) gene variants and alzheimer's disease, affective disorders, posttraumatic stress disorder, schizophrenia, and substance dependence. *Am J Med Genet B Neuropsychiatr Genet,* 2006 141B, 387-393.
[40] Sun, RF; Zhu, YS; Kuang, WJ; Liu, Y; Li, SB. The G-712A polymorphism of brain-derived neurotrophic factor is associated with major depression but not schizophrenia. *Neurosci Lett,* 2011 489, 34-37.

[41] Pezet, S; Malcangio, M. Brain-derived neurotrophic factor as a drug target for CNS disorders. *Expert Opin Ther Targets,* 2004 8, 391-399.

In: Biochemistry and Molecular Biology
Editor: Mary Miu Yee Waye
ISBN: 978-1-63482-312-8

Chapter 5

A 1.36-Megabase Deletion in Chromosome 18q21.32 Including *GRP* in a Patient with Epilepsy, Borderline Intellectual Disability, and Multiple Cerebral Cavernous Malformations

***Larry Baum**[*1], **Youling Guo**[2], **Hongsheng Gui**[2], **Pak Chung Sham**[2], **Stacey S. Cherny**[2] and **Patrick Kwan**[3,4,5]*

[1]School of Pharmacy, The Chinese University of Hong Kong, Shatin, Hong Kong, China
[2]Department of Psychiatry and The State Key Laboratory of Brain and Cognitive Sciences, The University of Hong Kong, Pokfulam, Hong Kong, China
[3]Division of Neurology, Department of Medicine and Therapeutics, Prince of Wales Hospital, The Chinese University of Hong Kong, Shatin, Hong Kong, China
[4]Department of Medicine, The University of Melbourne, Royal Melbourne Hospital, Melbourne, Australia
[5]Department of Neurology, Royal Melbourne Hospital, Melbourne, Australia

Abstract

From analysis of copy number variations (CNVs) in a genome-wide association study (GWAS) of symptomatic epilepsy, we discovered a 1.36 megabase deletion at chromosome 18q21.32 in a patient with borderline intellectual disability, multiple cerebral cavernous malformations (CMs), and complex partial seizures from age one year. The deletion was confirmed by quantitative polymerase chain reaction. The deleted region included the genes *ALPK2*, *MALT1*, *ZNF532*, *GRP*, *RAX*, *CPLX4*, *LMAN1*, and *CCBE1*. GRP (gastrin-releasing peptide, or bombesin) and GRP receptor are expressed in the hippocampus, where GRP depolarizes GABAergic interneurons, thus inhibiting CA1

[*] Correspondence to: Prof. Larry Baum. School of Pharmacy. The Chinese University of Hong Kong. Shatin, Hong Kong, China. Phone: 852-39436833. Fax: 852-26035295. E-mail: lwbaum@cuhk.edu.hk

pyramidal neurons and suppressing seizures. GRP receptor gene disruption has previously been reported in a patient with epilepsy and low intelligence. We speculate that the combined effects of multiple CMs and decreased GRP may have induced epilepsy in this patient. The finding of both genetic and structural abnormalities that could have resulted in epilepsy independently and interactively in this patient illustrates the challenge in classifying epilepsy syndromes.

Keywords: Epilepsy, genetics, copy number variant, 18q deletion, cerebral cavernous malformation, cavernoma, GRP, bombesin

Introduction

While idiopathic epilepsy has proven a fruitful hunting ground for discovering epilepsy genes, genetics in symptomatic or structural/metabolic syndrome has remained less explored because brain insults are, by definition, the "cause" of the epilepsy. However, given that only some patients suffering brain insults develop epilepsy, including about 15% annually of people with multiple intracranial cavernous malformations [1], other factors including genetics may influence the susceptibility to lesional epilepsy.

We therefore conducted a genome-wide association study (GWAS) of symptomatic epilepsy, reporting single nucleotide polymorphisms (SNPs) associated with disease risk at genome-wide significance level [2].

In addition to SNPs, copy number variants (CNVs) in GWAS may provide valuable clues to discovering genes playing a role in disease. We aimed to locate variants as contributing causes, in addition to other genetic or lesional factors. Of the two longest CNVs in our GWAS that included genes and were only present in the patients but not controls, we report the details of one of these CNVs in 18q21.32.

Methods

The study was approved by the ethics committees of the participating hospitals, and all subjects or their guardians gave written informed consent. CNVs were detected in a subset of the GWAS subjects, 483 Hong Kong Chinese patients with symptomatic focal epilepsy (adapted from the revised international organization of phenotypes in epilepsy [3]) and 845 non-epilepsy controls (101 from a hypertension study and 797 with low or high quartiles of bone mineral density from an osteoporosis study, with 53 removed after quality control). Briefly, as a quality control measure, we removed samples showing low quality in SNP genotyping and samples prone to bias in CNV detection [4]. Next, CNVs were detected by two programs, PennCNV (http://www.openbioinformatics.org/penncnv/) and QuantiSNP (http://www.well.ox.ac.uk/quantisnp), and were then filtered to remove abnormal CNVs [5;6]. We imputed genotypes of SNPs within a 50 kilobase region centered on *GRP* using Impute2 software, using genotypes from the 1000 Genomes Project as the reference panel [7]. Qualified genotypes of the above SNPs (missing rate < 0.05 and imputation information score > 0.8) were tested for association with epilepsy by SNPtest 2.0 [8], adopting an allelic model and 15 principle components as covariates.

To confirm the extent of the 18q21.32 deletion identified, quantitative polymerase chain reaction (qPCR) was performed in *GRP*, *ALPK2*, and *CCBE1*. Primers were designed for quantitative PCR (qPCR) using Primer3 software (http://frodo.wi.mit.edu/) [9] and the following parameters:

- Primer length: 9-30 base pairs (bp), with an optimum of 20 bp
- Primer melting temperature (T_m): 58-60°C, with an optimum of 59°C
- Maximum T_m difference between primer pairs: 2°C
- GC content of primers: 30-80%, with an optimum of 50%
- Amplicon length: 80-150 bp
- No more than two G or C bases among the 5 nucleotides at the 3'-end
- Maximum 3' stability value: ≥3 or more, with an optimum of 3
- No runs of ≥4 identical nucleotides (especially G bases).
- No complementarity of last 4 nucleotides at the 3' ends of primers to any other sequence in either primer

Secondary structure of amplicons was checked using Quikfold software (http://mfold.rna.albany.edu/?q=DINAMelt/Quickfold) [10;11] and the following parameters:

- Theoretical T_m of primers
- 50 mM Na^+
- 1.5 mM Mg^{2+}

Amplicons with secondary structure, indicated by $\Delta G<0$, were avoided (especially for secondary structures at primer annealing sites). Primers containing SNPs were avoided. Primers predicted by Primer-BLAST (http://www.ncbi.nlm.nih.gov/tools/primer-blast/) to amplify sequences other than the desired amplicon were avoided. Reference primers in a region without CNVs (the *GAPDH* gene) were chosen from RTprimerDB (http://rtprimerdb.org/) [12].

Primers were as follows:

GRP: TGGTGCACACAGAATTGATG, CACCTCTGCCAAGTCGTTTA
ALPK2: TTTCAGCATCCTCGCATAAC, GATGTGCAGCAAAGGCTCTA
CCBE1: CTGGGTTGGGATTGTCTCTT, GTGGGCCAGATAGGGACTTA
GAPDH: GCTTGCCCTGTCCAGTTAAT, TAGCTCAGCTGCACCCTTTA

PCR was performed using SYBR Premix Ex Taq (Tli RNaseH Plus) PCR Master Mix (TaKaRa). The PCR reaction mixture contained 5 μL 2 × Master Mix, 0.2 μL ROX reference dye, 200 nM forward primer, 200 nM reverse primer, 4.4 ng genomic DNA template, and water up to 10 μL final volume. All samples were tested in triplicate. Standard curves for selected genes and reference gene amplification were prepared by a $\log_2$ serial dilution of a control patient DNA (calibrator) from 8 ng/μL to 0.25 ng/μL. Amplification was performed using an ABI Prism 7900 HT Sequence Detection System at 95°C for 30 s followed by 40 cycles of 95°C for 5 s and 60°C for 30s.

Clipped data generated by SDS version 2.3 were exported in text format, and Ct values were calculated using Real-time PCRMiner version 4.0 (http://www.miner.ewindup.info) [13]. Further DNA quantification was calculated using Microsoft Excel following the comparative standard curve method (ΔC_T). Briefly, values calculated from the standard curve of selected genes were normalized to that of the reference gene (gene of interest / reference gene). Then a second normalization was performed to calculate the relative expression to a control patient DNA (calibrator). Thus, relative expression values of 0.5, 1, 1.5, and 2 indicated copy numbers of 1, 2, 3, and 4, respectively.

RESULTS

A deletion of one copy of a 1.36 megabase portion of chromosome 18q21.32 was detected in one patient. Figure 1 displays the deleted region, which includes the RefSeq genes *ALPK2*, *MALT1*, *ZNF532*, *OACYLP* (pseudogene), *SEC11C* (provisional), *GRP*, *RAX*, *CPLX4*, *LMAN1*, and *CCBE1*. All except *ALPK2*, *ZNF532*, *OACYLP*, and *SEC11C* are listed in OMIM (omim.org).

The deletion contained 434 SNPs from the GWAS. Other CNVs in this patient were much shorter and were mostly intergenic or highly polymorphic in the population, and thus may be unlikely to affect phenotype. Among the 30 CNVs in this patient with length > one kb, quality score > 10, and > three probes, only five were longer than 100 kb (the second longest being about 230 kb), of which none other than the 1.36 Mb deletion were predicted to affect expression of a gene (as judged by duplication of an entire gene or deletion of an exon) except a chromosome 17 deletion of about 140 kb including part of ARL17; however, this gene has a copy, thus a deletion of one copy may decrease the gene dose by only 1/4. No other patients or controls in the GWAS had CNVs overlapping with *GRP*.

qPCR was performed near the middle and either end of the region, within *GRP*, *ALPK2*, and *CCBE1*. The patient with the deletion had from 0.98 to 1.34 copies, while the two GWAS patients displaying no CNV in the region had from 2.00 to 2.32 copies (Figure 1, bottom), thus confirming the extent of the deletion,

To explore the potential association of *GRP* with risk of epilepsy, genotypes of SNPs within a 50 kilobase region centered on *GRP* were imputed for GWAS subjects. Of the 482 SNPs either genotyped in the GWAS or imputed, none was significantly associated with epilepsy after Bonferroni correction for multiple comparisons. The lowest uncorrected *p*-value was 0.0033, in rs633083, which is located 46 kb 3' from *GRP* and 4 kb 5' from *RAX*.

The patient, currently aged 22 years, is an ethnic Han Chinese male from Hong Kong. He dropped out of the first year of high school. In spite of borderline intellectual disability he holds full-time employment as a chef. He is unmarried and has no children. There is no family history of epilepsy, intellectual disability, or other neurological abnormalities. He has drug-resistant epilepsy with complex partial seizures from age one year. His current seizure frequency is one per month despite treatment with phenytoin 300 mg/day and sodium valproate 1600 mg/day.

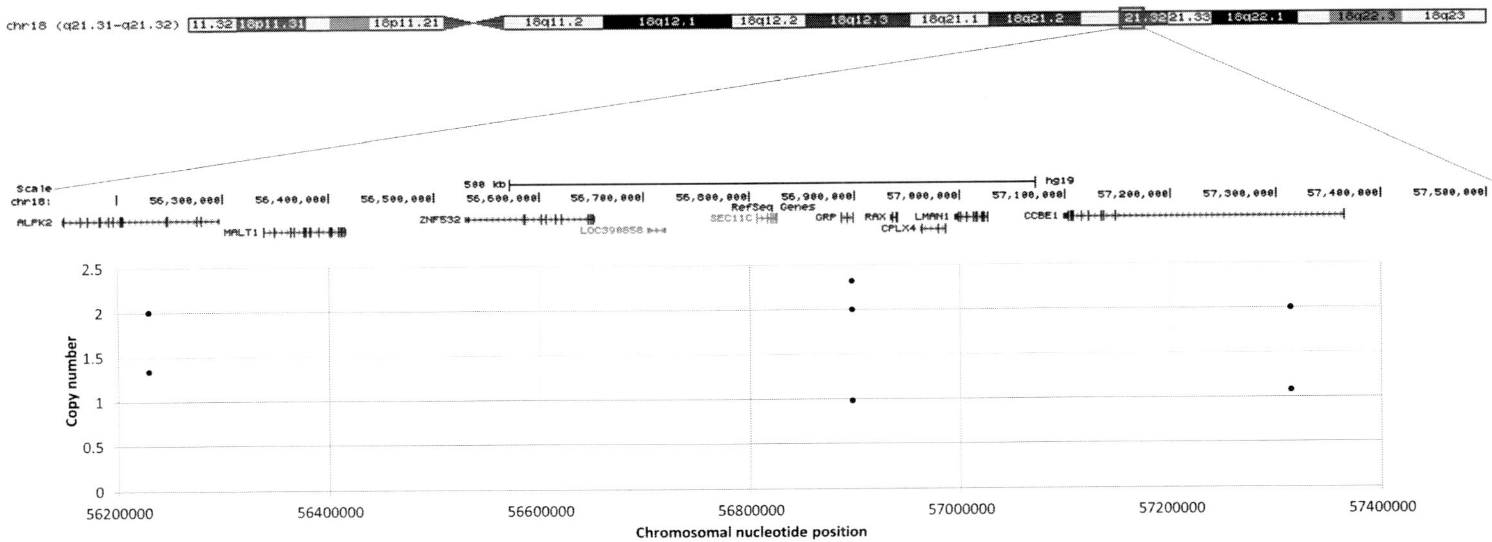

Figure 1. The deletion in chromosome 18q21.32 extends from nucleotide 56142499 to 57502401 (UCSC hg19 assembly) and includes the genes shown. The lower portion of the figure displays quantitative polymerase chain reaction measurements. The dots indicate the copy number of DNA sequences at three locations in the deletion, for the patient with the deletion (all the dots <1.5 copies) and for two patients with normal copy number (all the dots >1.5 copies).

Previous trial of lamotrigine was stopped after skin rashes developed, and carbamazepine was withdrawn due to significant neutropenia. Seizure semiology as captured on video EEG recording showed right hand dystonia and impaired awareness, accompanied with ictal discharge in bi-frontal leads. The first brain MRI of the patient was performed at age five years and showed multiple CMs. The latest brain MRI, in 2007, showed over 20 CMs distributed in both cerebral hemispheres. The largest CM was in left frontal lobe.

DISCUSSION

We have reported a patient with a 1.36 Mb deletion at chromosome 18q21.32, multiple CMs, epilepsy and borderline intellectual disability. Patients with deletion of part of the long arm of chromosome 18, or 18q deletion syndrome (18q-), are known to have some or all of a variety of phenotypes, including mental retardation, short stature, hypotonia, hearing impairment, and foot deformities (OMIM 601808). Most of 18q21 was deleted in two patients with microcephaly and severe mental retardation but without epilepsy [14]. Epilepsy has been reported in 10-31% of 18q- patients, with seizures mainly focal, beginning in the first years of life, and controllable by drugs [15]. The portion of chromosome 18q critical for epilepsy has not been determined, perhaps because of the partial penetrance, but usually involves genes more distal than the deletion in our patient [15].

Although CMs are recognized as a factor contributing to epilepsy, some sense of the extent of the contribution could be ascertained from a prospective population-based study that followed patients with CMs without history of hemorrhage. Five years after CM diagnosis, only 4% of patients developed a first-ever seizure [1], with the incidence increasing to 75% in those with multiple CMs [1]. This patient developed epilepsy at one year of age. CMs can result in epilepsy in children this young [16]. As far as we are aware, no case of 18q deletion syndrome with concurrent CMs has been reported previously.

It is possible that the genes in the deleted region might have played a role in the pathogenesis of CMs and/or epilepsy in this patient. For instance, recessive *RAX* mutations cause anophthalmia and cerebral defects [17]. However, heterozygous *RAX* mutations were not reported to alter phenotypes, though perhaps asymptomatic abnormalities, such as CMs, were undetected. Recessive mutations of other genes in the deleted region have been reported to cause diseases, such as *LMAN1* in patients with combined factor V and factor VIII deficiency [18] and *CCBE1* in patients with Hennekam lymphangiectasia-lymphedema syndrome (OMIM 235510), however the lack of similar phenotypes in our patient suggests that loss of one copy of these genes is not critical. Since CMs have not been reported in patients with 18q- syndrome, CMs in this patient may not be caused by the deletion. In this patient, we detected no CNVs in the reported CM genes, *CCM1*, *CCM2*, and *PDCD10* (OMIM 116860), though we did not perform sequencing to detect small mutations. About half of CM cases are familial, thus a mutation in a CM gene may exist in this patient [19].

Another plausible candidate is *GRP*. GRP (gastrin-releasing peptide, or bombesin) is found in gut, lungs, and brain. In rats, it binds with high affinity to the GRP receptor, or Bombesin BB_2 receptor, and mediates diverse functions, including secretion of many gut hormones and control of temperature, eating, and grooming [20-22]. In rat brain, GRP binding sites are concentrated in the supraoptic nucleus, arcuate nucleus, and outer rim of the hippocampal oriens layer [20]. When applied to rat hippocampal slices, GRP depolarized GABAergic interneurons in the stratum oriens [21] and depressed epileptiform activity [23]. By contrast, in CA1 pyramidal neurons, GRP had minimal effect on resting membrane properties, though it did greatly increase the number and amplitude of miniature inhibitory postsynaptic currents [21]. These observations suggest that GRP inhibits interneurons that synapse on pyramidal neurons, thus inhibiting them.

Applying GRP to rats using *in vivo* hippocampal microdialysis increased GABA levels in the hippocampus, and intracerebroventricular administration of GRP to audiogenic seizure-prone DBA/2 mice suppressed seizures, increasing the number of mice convulsing and the latency to tonic seizures, and decreasing the incidence of lethality [24]. A patient with epilepsy, severe intellectual disability, autism, and multiple exostoses has been reported to have a chromosome X;8 translocation within the GRP receptor (*GRPR*) gene [25]. Thus, decreased GRP function may increase propensity to develop seizures.

Since only one copy is deleted in our patient, we may consider whether haploinsufficiency of *GRP* is sufficient to induce epilepsy, and whether the coexistence with multiple CMs is a mere coincidence. Many 18q- patients whose deletion includes *GRP* do not have epilepsy (seven out of 11 patients in one series [14]), therefore the penetrance of *GRP* deletions for causing epilepsy cannot be high [26;27] (https://www.iscaconsortium.org/).

Alternatively, in our patient, a reduced level of GRP combined with structural brain abnormality, such as CMs, might have pushed the patient over a threshold leading to epilepsy. In a precedent for such a combination of risk factors perhaps causing epilepsy, two CNVs, co-existing deletions of 16p11.2 and 19p13.2, were suggested to contribute to epilepsy [28]. To test our hypothesis, further studies may include comparing the prevalence of epilepsy among 18q- patients with and without *GRP* deletion, and use of GWAS data to compare the prevalence of *GRP* deletion between epilepsy patients and the general population. The latter will likely require examination of very large cohorts through meta-analysis of multiple epilepsy GWAS, an effort that is being undertaken by the ILAE Consortium on Genetics of Complex Epilepsies.

An alternative means to demonstrate an association between epilepsy and genetic variants in *GRP* could be to analyze genotypes of SNPs in or near *GRP*. In our GWAS, SNPs within a 50 kilobase region centered on *GRP* were not significantly associated with epilepsy. There are several possible implications: GRP variants do not affect epilepsy risk; no variants in GRP affect its expression or function; or our GWAS is too small to detect a significant association.

The finding of both genetic and structural abnormalities that could have resulted in epilepsy independently and interactively in this patient illustrates the challenge in the current approach to classify epilepsy into mutually exclusive syndromes such as "genetic" or "structural/metabolic" [3].

ACKNOWLEDGMENTS

We thank patients who participated in this study. We thank Mr. Raymond WK Cheung and Dr. Ying Kit Cheung for assisting with qPCR. We also thank Professor Annie W.C. Kung, Department of Medicine, The University of Hong Kong, for contributing to GWAS control data. Funding was provided by the Research Grants Council of the Hong Kong

Special Administrative Region, China (project numbers HKU7623/08M, HKU7747/07M and CUHK4466/06M) and Chinese University of Hong Kong Direct Grant for Research 2011.1.096.

REFERENCES

[1] Josephson, CB; Leach, JP; Duncan, R; Roberts, RC; Counsell, CE; Al-Shahi Salman, R; Scottish Audit of Intracranial Vascular Malformations (SAIVMs) steering committee and collaborators. Seizure risk from cavernous or arteriovenous malformations: Prospective population-based study. *Neurology,* 2011 76, 1548-1554.

[2] Guo, Y; Baum, LW; Sham, PC; Wong, V; Ng, PW; Lui, CH; Sin, NC; Tsoi, TH; Tang, CS; Kwan, JS; Yip, BH; Xiao, SM; Thomas, GN; Lau, YL; Yang, W; Cherny, SS; Kwan, P. Two-stage genome-wide association study identifies variants in CAMSAP1L1 as susceptibility loci for epilepsy in chinese. *Hum Mol Genet,* 2012 21, 1184-1189.

[3] Berg, AT; Berkovic, SF; Brodie, MJ; Buchhalter, J; Cross, JH; van Emde, BW; Engel, J; French, J; Glauser, TA; Mathern, GW; Moshe, SL; Nordli, D; Plouin, P; Scheffer, IE. Revised terminology and concepts for organization of seizures and epilepsies: Report of the ILAE commission on classification and terminology, 2005-2009. *Epilepsia,* 2010 51, 676-685.

[4] Pugh, TJ; Delaney, AD; Farnoud, N; Flibotte, S; Griffith, M; Li, HI; Qian, H; Farinha, P; Gascoyne, RD; Marra, MA. Impact of whole genome amplification on analysis of copy number variants. *Nucleic Acids Res,* 2008 36, e80.

[5] Wang, K; Li, M; Hadley, D; Liu, R; Glessner, J; Grant, SF; Hakonarson, H; Bucan, M. PennCNV: An integrated hidden markov model designed for high-resolution copy number variation detection in whole-genome SNP genotyping data. *Genome Res,* 2007 17, 1665-1674.

[6] Colella, S; Yau, C; Taylor, JM; Mirza, G; Butler, H; Clouston, P; Bassett, AS; Seller, A; Holmes, CC; Ragoussis, J. QuantiSNP: An objective bayes hidden-markov model to detect and accurately map copy number variation using SNP genotyping data. *Nucleic Acids Res,* 2007 35, 2013-2025.

[7] Howie, B; Marchini, J; Stephens, M. Genotype imputation with thousands of genomes. *G3.(Bethesda.),* 2011 1, 457-470.

[8] Marchini, J; Howie, B. Genotype imputation for genome-wide association studies. *Nat Rev Genet,* 2010 11, 499-511.

[9] Rozen, S; Skaletsky, H. Primer3 on the WWW for general users and for biologist programmers. *Methods Mol Biol,* 2000 132, 365-386.

[10] Markham, NR; Zuker, M. UNAFold: Software for nucleic acid folding and hybridization. *Methods Mol Biol,* 2008 453, 3-31.

[11] Markham, NR; Zuker, M. DINAMelt web server for nucleic acid melting prediction. *Nucleic Acids Res,* 2005 33, W577-W581.

[12] Lefever, S; Vandesompele, J; Speleman, F; Pattyn, F. RTPrimerDB: The portal for real-time PCR primers and probes. *Nucleic Acids Res,* 2009 37, D942-D945.

[13] Zhao, S; Fernald, RD. Comprehensive algorithm for quantitative real-time polymerase chain reaction. *J Comput Biol,* 2005 12, 1047-1064.

[14] Feenstra, I; Vissers, LE; Orsel, M; van Kessel, AG; Brunner, HG; Veltman, JA; van Ravenswaaij-Arts, CM. Genotype-phenotype mapping of chromosome 18q deletions by high-resolution array CGH: An update of the phenotypic map. *Am J Med Genet A,* 2007 143A, 1858-1867.

[15] Grosso, S; Pucci, L; Di Bartolo, RM; Gobbi, G; Bartalini, G; Anichini, C; Scarinci, R; Balestri, M; Farnetani, MA; Cioni, M; Morgese, G; Balestri, P. Chromosome 18 aberrations and epilepsy: A review. *Am J Med Genet A,* 2005 134A, 88-94.

[16] Bigi, S; Capone Mori, A; Steinlin, M; Remonda, L; Landolt, H; Boltshauser, E. Cavernous malformations of the central nervous system in children: Presentation, treatment and outcome of 20 cases. *Eur J Paediatr Neurol,* 2011 15, 109-116.

[17] Abouzeid, H; Youssef, MA; Bayoumi, N; Elshakankiri, N; Marzouk, I; Hauser, P; Schorderet, DF. RAX and anophthalmia in humans: Evidence of brain anomalies. *Mol Vis,* 2012 18, 1449-1456.

[18] Zhang, B; Spreafico, M; Zheng, C; Yang, A; Platzer, P; Callaghan, MU; Avci, Z; Ozbek, N; Mahlangu, J; Haw, T; Kaufman, RJ; Marchant, K; Tuddenham, EG; Seligsohn, U; Peyvandi, F; Ginsburg, D. Genotype-phenotype correlation in combined deficiency of factor V and factor VIII. *Blood,* 2008 111, 5592-5600.

[19] Cavalcanti, DD; Kalani, MY; Martirosyan, NL; Eales, J; Spetzler, RF; Preul, MC. Cerebral cavernous malformations: From genes to proteins to disease. *J Neurosurg,* 2012 116, 122-132.

[20] Ladenheim, EE; Jensen, RT; Mantey, SA; McHugh, PR; Moran, TH. Receptor heterogeneity for bombesin-like peptides in the rat central nervous system. *Brain Res,* 1990 537, 233-240.

[21] Lee, K; Dixon, AK; Gonzalez, I; Stevens, EB; McNulty, S; Oles, R; Richardson, PJ; Pinnock, RD; Singh, L. Bombesin-like peptides depolarize rat hippocampal interneurones through interaction with subtype 2 bombesin receptors. *J Physiol (Lond),* 1999 518 (Pt 3), 791-802.

[22] Spindel, ER; Zilberberg, MD; Habener, JF; Chin, WW. Two prohormones for gastrin-releasing peptide are encoded by two mRNAs differing by 19 nucleotides. *Proc Natl Acad Sci U S A,* 1986 83, 19-23.

[23] Zhang, HP; Xiao, Z; Cilz, NI; Hu, B; Dong, H; Lei, S. Bombesin facilitates GABAergic transmission and depresses epileptiform activity in the entorhinal cortex. *Hippocampus,* 2014 24, 21-31.

[24] Andrews, N; Davis, B; Gonzalez, MI; Oles, R; Singh, L; McKnight, AT. Effect of gastrin-releasing peptide on rat hippocampal extracellular GABA levels and seizures in the audiogenic seizure-prone DBA/2 mouse. *Brain Res,* 2000 859, 386-389.

[25] Ishikawa-Brush, Y; Powell, JF; Bolton, P; Miller, AP; Francis, F; Willard, HF; Lehrach, H; Monaco, AP. Autism and multiple exostoses associated with an X;8 translocation occurring within the GRPR gene and 3' to the SDC2 gene. *Hum Mol Genet,* 1997 6, 1241-1250.

[26] Firth, HV; Richards, SM; Bevan, AP; Clayton, S; Corpas, M; Rajan, D; Van Vooren, S; Moreau, Y; Pettett, RM; Carter, NP. DECIPHER: Database of chromosomal imbalance and phenotype in humans using ensembl resources. *Am J Hum Genet,* 2009 84, 524-533.

[27] International standards for cytogenetic arrays consortium. all known benign copy-number variants in the consortium's database (https://www.iscaconsortium.org/images/stories/isca/ISCA_Known_benign_regions_hg18.txt).

[28] Bassuk, AG; Geraghty, E; Wu, S; Mullen, SA; Berkovic, SF; Scheffer, IE; Mefford, HC. Deletions of 16p11.2 and 19p13.2 in a family with intellectual disability and generalized epilepsy. *Am J Med Genet A,* 2013 161A, 1722-1725.

Reviews

In: Biochemistry and Molecular Biology
Editor: Mary Miu Yee Waye
ISBN: 978-1-63482-312-8

Chapter 6

Analysis of Codon Bias and Its Application on the Expression of Foreign Genes in Fungi

Ruan Ban-Zhan and Xie Bao-Gui*
Mycological Research Center of Fujian Agricultural and Forestry University, Fuzhou, China

Abstract

Codon bias refers to the nonrandom usage of synonymous codons for encoding amino acids in organisms. As it is related to translation efficiency, this phenomenon implicates important biological sense. Fungi are good hosts for the expression of foreign genes, because of its fast growth and reproduction, posttranslational processing and closer homology with higher eukaryotic organisms (plants and vertebrates). Currently, researchers have encountered low expression level in many fungi transgenic research, so it is of great practical significance and economical value to investigate the application of codon bias analysis in foreign gene expression in fungi. In this review, we summarize the basic theories and analysis methods about codon bias, and introduce the application of codon bias analysis in foreign gene expression in three categories of fungi - yeasts, molds and mushrooms respectively.

Keywords: Codon bias; gene expression; fungi

Introduction

Codons are basic rules of the information carried by proteins corresponding with the information carried by nucleic acids. It is also a basic link of the transmission of information transfer in organisms. The fact that genetic codons are composed of three nucleotides was found firstly by a mutagenesis experiment with T4 phage [1]. Nirenberg et al. firstly identified that the codon encoding phenylalanine was UUU by an mRNA artificially

* E-mail address: mrcfafu@163.com.

synthesized and detecting the protein translated *in vitro* [2]. Through the tireless efforts of scientists and the improvement of deciphering method for genetic codes, it took 5 years to complete the deciphering of all the genetic codes. There are 61 codons in all encoding the 20 amino acids of natural proteins, and each amino acid corresponds to 1-6 codons. The codons encoding the same amino acid are called synonymous codons for each other. In most organisms, synonymous codons are not selected to be used in equal frequencies [3]. When it comes to a particular species of an organism or a gene, certain synonymous codons tend to be used preferentially as the protein coding sequences. This phenomenon is called synonymous codon usage bias [4]. For a species of an organism, the phenomenon of codon preference is more significant in its highly expressed genes than lowly expressed genes [5]. There are several hypotheses on the cause of this phenomenon and the most widely recognized one is the hypothesis "Mutation-selection Balance" [6]. In the opinion of this hypothesis, selection pressure leads to the tendency of organisms to encode amino acids with the most optimal codons, and the presence of non-optimal codons is due to mutation. This selection pressure is mainly reflected in the efficiency of translation, and it is especially evident on the highly expressed genes [7]. The dynamic balance of these two forces gradually results in the formation of the codon bias diversity in different species' genomes. Although there is not very conclusive evidence for the hypothesis "Mutation-selection Balance", attention has been shifted to the relationship between codon bias and genes expression, as well as the study of the translational regulation mechanism of codon bias. With the development of Bioinformatics and the increasingly convenience of getting oligonucleotides by chemical synthesis, it becomes possible to put analysis software and online analysis site into marvelous codons usage analysis. Moreover, it also becomes a reality that the sequence of several kilobytes' oligonucleotides are reconstructed, then artificially synthesized and transformed into the genomes of other organisms. In this way, the foreign genes may be expressed in different hosts' cells and the expression level of those exogenous proteins may be increased. This is a brief overview of the basic theories and analysis methods about codon bias, as well as the application of codon bias analysis in foreign genes expression in three categories of fungi, including yeasts, molds and mushrooms.

Basic Theories and Analysis Methods about Codon Bias

With the emergence and development of bioinformatics, codon usage analysis softwares are constantly updated, showing a lot of characteristics, like easy operation, high accuracy. Some websites supporting online analysis are also greatly convenient for researchers to choose some appropriate methods to analyze the features of codon usage.

Relative Synonymous Codon Usage (RSCU)

An RSCU value for a codon refers to the ratio between the actual frequency of codon usage of a particular codon and its expected frequency in a case when all codons are equally used. It is calculated as follows:

$$\text{RSCU}_{ij} = \frac{x_{ij}}{\frac{1}{n_i}\sum_{j=1}^{n_i} x_{ij}}$$

where x_{ij} is the number of occurrences of the jth codon for the ith amino acid, and n_i is the number (from one to six) of alternative codons for the ith amino acid [8].

CODON ADAPTATION INDEX (CAI)

The CAI uses a reference set of highly expressed genes from a species to assess the relative merits of each codon, and assesses the proximity of a gene's codon preference and these highly expressed genes' preference [9]. The relative adaptiveness of a codon, W_{ij}, is then the frequency of use of that codon compared to the frequency of the optimal codon for that amino acid. W_{ij} and CAI are calculated as follows:

$$W_{ij} = \frac{\text{RSCU}_{ij}}{\text{RSCU}_{i\max}} = \frac{X_{ij}}{X_{i\max}}$$

$$\text{CAI} = \left(\prod_{K=1}^{L} W_K\right)^{\frac{1}{L}} \quad 1$$

where $\text{RSCU}_{i\max}$ and $X_{i\max}$ are the RSCU and X values for the most frequently used codon for the ith amino acid, L is the number of codons in the gene, and W_k is the weight of the kth codon in the gene sequence.

The CAI value ranges between 0 and 1. It was found that the larger the value, the more highly frequent the codon. CAI is generally used to predict the expression level of intraspecies genes. For example, this method was applied in the codon usage analysis of the Boophius microplus bm86 gene [10] and *Deinococcus* radiodurans [11].

EFFECTIVE NUMBER OF CODONS (ENC OR NC) [12, 13]

The ENC introduced by Frank Wright has been widely used to measure codon usage evenness. It is calculated as follows:

$$F = \frac{n\sum_{i=1}^{k} p_i^2 - 1}{n-1} \quad n > 1 \quad p_i = \frac{n_i}{n} \quad \text{ENC} = 2 + \frac{9}{F_2} + \frac{1}{F_3} + \frac{5}{F_4} + \frac{3}{F_6}$$

where P_i denotes the frequency of the ith synonymous codon, k is the number of synonymous codons of an amino acid, and n is the number of the translated amino acid in a gene. Usually, highly expressed genes' codons tend to use a specific one or several synonymous codons. It was found that the smaller the ENC value, the larger the preference degree. Conversely, in lowly expressed genes, there is a greater variety of rare codons, the preference degree is lesser, and the ENC value is bigger.

Therefore, the ENC value can be used to determine the amount of the expression levels of endogenous genes. However, it will be affected by the composition and length of amino acids. This method was applied in the codon usage analysis of *Escherichia coli* [14]、Human Bocavirus [15].

Codon Preference Parameter (CPP) [16]

The CPP reflects what extent a group of codons are being used nonrandomly. It can be calculated using a formula as follows:

$$\mathrm{CPP} = \sum_{i}^{18} \frac{\left| x_{ij} - \dfrac{\sum_{j=1}^{n_i} x_{ij}}{n_i} \right|}{\sum_{j=1}^{n_i} x_{ij}} \times \frac{n_i}{(2n_i - 2)}$$

where xij is the number of occurrences of the jth codon for the ith amino acid and ni is the number of alternative codons for the ith amino acid. It will not be taken into consideration if those amino acids has no alternative codons, that is to say, ni is not equal to 1. An essentially random distribution would result in a CPP of 0 and a totally nonrandom distribution would result in a value of 18.

The CPP can be used to compare not only codon usage in different coding regions but also codon usage patterns of different organisms. The above two purposes are not easily subjected to the influence of overall amino acid composition of the gene product and the collection of gene products contained in the sum. This makes comparisons within and between species more meaningful.

Frequency of Optimal Codons (FOP) [17]

The optimal codon of an amino acid refers to the one that is mostly used in highly expressed genes. It is also defined as the codon with the highest number of tRNA genes for its anticodon, among its synonymous codons. The FOP has species specificity. A FOP is associated with a DNA sequence and its corresponding expression information.

The simplest way to compute the FOP of a gene is as follows:

$$\mathrm{FOP_s}(g) = \frac{1}{\mathrm{N}} \sum_i \mathrm{n}_i(g)$$

where the subscript s stands for "simple", $\mathrm{n}_i(g)$ is the count of the codon i in the gene g and N is the total number of codons in g. The FOP calculated through this method is easily subjected to the influence of the amino acid composition. In order to obtain a measure which is independent of amino acid composition, the above formula is optimized as follows:

$$\mathrm{FOP(g)} = \frac{1}{\mathrm{N}} \sum_i \mathrm{syn}(i)\mathrm{n}_i(g)$$

where syn(i) is the degeneracy of the amino acid coded by i. As this formula shows, a gene close to random synonymous codon usage will have a FOP value close to 1, regardless of its amino acid composition.

The above formula is further summarized as follows:

$$\mathrm{FOP}(g) = \sum_i f_{aa(i)}(g)\mathrm{RSCU}_i(g)$$

where

$$f_{aa(i)}(g) = \frac{\mathrm{n}_{aa(i)}(g)}{\mathrm{N}}$$

As shown in the above formula, the FOP becomes the weighted average of the optimal codons' RSCU value, whose weighted coefficient is the amino acid usage frequency.

CODON BIAS INDEX (CBI) [18]

The CBI is devised for measuring deviation from uniform use of synonymous codons in a codon use table, which is shown as follows:

$$\mathrm{CBI} = \sum_{i=1}^{18} \mathrm{n}_i / \mathrm{n}_{\mathrm{tot}} \left(\sum_{j=1}^{\mathrm{S}_i} (1 - \mathrm{R}_{ij})^2 / (\mathrm{S}_i - 1) \right)$$

where R value is the relative use of synonymous codons [9], and it can be calculated as follows:

$$\mathrm{R}_{ij} = \mathrm{n}_{ij} / \mathrm{n}_{i\,\mathrm{max}}$$

In the above two equations, Si is the number of synonymous codons for the ith amino acid, n_i is the number of occurrences of the ith amino acid, n_{tot} is the total number of residues excluding methionine (met) and tryptophan (trp) and since they have only one codon each, n_{ij} is the number of occurrences of the jth synonymous codon of amino acid i and n_{imax} is the number of occurrences of the most frequently used synonymous codon of the ith amino acid. The value of CBI ranges from 0 to 1.

Relative Frequency of Synonymous Codons (RFSC)

The RFSC is the ratio of the number of a codon observed in a sample to the total number of the corresponding amino acid in a sample. The value of RFSC can be calculated as follows:

$$RFSC_{ij} = \frac{x_{ij}}{\sum^{n_i} x_{ij}}$$

In the above formula, x_{ij} and n_i are the appearances of the jth codon and the species of its synonymous codons encoding the ith amino acid respectively. RFSC reflects the frequency of occurrence of each codon in a sample regardless of the influence of the length and the abundance of amino acids of a single gene of the sample. This analysis method has been widely used in recent years [19-21].

Relative Codon Adaptation (RCA) [22]

RCA follows the same principles applied to CAI and is an extension of the basic principle of relative codon frequencies. It makes use of a single given reference set to compute observed and expected codon frequencies and takes into account the genomic base compositions. Formally,

$$RCA = \left(\prod_{i=1}^{L} RCA_{xyz}(i) \right)^{1/L} \quad RCAxyz = \frac{\chi(xyz)}{\chi_1(x)\chi_2(y)\chi_3(z)}$$

where $\chi(xyz)$ is the observed frequency of codon xyz in any particular reference gene set, $\chi_n(m)$ the observed frequency of base m at codon position n in the same preference set, and L the length in codons of the query sequence. Like CAI, RCA is also computed as the geometric mean of the RCA_{xyz} term for each codon xyz in the sequence of interest.

What's more, RCA can be used in any reference sequence set without requiring genomic corrections or limiting their usage to sequences larger than 1,000 bp.

CODON DEVIATION COEFFICIENT (CDC) [23]

Relatively speaking, the CDC is a new analysis method for codon usage. In this method, both GC and purine contents are taken into consideration in sequences as well as at the three codon positions of the coding sequences.

What's more, cosine distance metric and bootstrapping are adopted, which makes it possible to quantify codon usage bias and ascertain its statistical significance. It can be calculated by the following formula.

$$CDC = 1 - \frac{\sum_{xyz} \pi_{xyz} \times \hat{\pi}_{xyz}}{\sqrt{\sum_{xyz} \pi_{xyz}^{2} \times \sum_{xyz} \hat{\pi}_{xyz}^{2}}}$$

where x, y, z ∈ {A, T, G, C}, the expected usage π_{xyz} is defined as the product of its constituent expected nucleotide contents $x_1y_2z_3$, i.e., all the sense codons are normalized, while $\hat{\pi}_{xyz}$ refers to the observed usage. The CDC renders a distance coefficient ranging from 0 (no bias) to 1 (maximum bias).

The CDC has a wide range of uses. For those genes or genomes with a variety of sequence composition, the CDC can be used to determine comparative magnitudes and patterns of optimal codon usage.

As the methods described above have their own characteristics, they can be applied to different situations.

It has become a conventional idea to explore in this field that integrating the features of DNA sequences and the biological functions of the encoded products by using the analysis methods of codon bias usage, then explaining the biological significance through the application of correlation analysis and clustering analysis methods [24].

THE APPLICATION OF CODON BIAS ANALYSES ON FOREIGN GENES EXPRESSION IN FUNGI

Since codons' degeneracy, different species have different bias in codon usage, leading to low [20] even no [25] expression of foreign genes in hosts' cells.

Many factors may have an impact on codon usage bias, such as the expression level of genes [26], translation efficiency [27], the nucleotide sequence of genes, dinucleotide frequency [28], G + C content [29], the length of genes [30-33], the abundance of tRNA [34], the structure and function of encoded proteins [35], the secondary structure of mRNA [36] and codon-anticodon binding strength etc. Of all the above factors, the key to decide whether codons adapt to the highly expressed genes is the abundance of tRNA [37].

Although the main factors and to what extent these factors affecting codon preference are still unknown, many of the previous studies show that using the advanced bioinformatics technology to analyze the codon usage of the expression systems and finding out their codon preference, then replacing the rare codons of foreign gene with the preferred codons of the hosts' genomes could lead to optimization of protein expression.

Finally, the foreign genes whose codons have been optimized are put into the hosts' cells, and genes can then be expressed freely in the expression systems of interest. Through this way, the expression level of foreign genes can be improved.

In recent years, with the continuous development of molecular biology and genetic engineering, there are many reports about improving the expression efficiency of foreign genes in host organisms by the method described above. Diverse host species were reported throughout animals [38], plants [39] and microorganisms [40-47]. In the existing reports, genes expressed in *E. Coli* are the most common, including human genes [40-42], animal genes [43, 44], plant genes [45], and microbial genes [46, 47].

This review, however, is focused on the application of this strategy in fungi as host organisms, and it could be expanded into yeasts, molds and mushrooms. At the same time, the research status and problems encountered in each category of fungi will be presented with relevant examples.

The Application in Yeasts

Yeasts, as common eukaryotic microorganisms, are a special category of single-celled fungi for transgenic research. They are usually selected as hosts for expressing foreign genes. There have been many reports about improving the expression level of foreign genes in yeasts by codon optimization in recent years. There are, in principle, two strategies to optimize the gene: optimize the consensus sequence around the starting ATG to ensure efficient translation initiation as well as optimize the genetic codons to be host specific [48].

Oberq F et al. mimicked the yeast consensus sequence (ATGTCT) of hAQP1, and they found that the sequence apparently has a negative influence on the production level [49]. Through the comparison of the expression of several Tn5-derived neomycin resistance genes whose seven additional ATG codons in the 5'-untranslated region were completely or partially removed, Yaqi S et al. found that the expression of Tn5-derived neomycin resistance gene can be enhanced by the removal of upstream ATG sequences [50].

Compared with the above strategy, another one has been applied more widely. Tu Y et al. achieved highly expressed recombinant porcine circovirus type 2 capsid protein in *Pichia pastoris* through codon optimization [51]. Luang Z [52] and Luang S [53] also obtained high-level expression of foreign proteins in *Pichia pastoris*.

The Application in Molds

As filamentous fungi, molds are widely distributed in nature. They are commonly found in the deterioration of cheese and fruit. Codon bias analyses has also been applied in exogenous genes expression in molds, whose top common genera are widely diverse, include *Aspergillus*, *Mucor, Rhizopus,* and *Penicillium* etc. Because of molds' own factors, they are rarely selected as hosts to express exogenous proteins. Meanwhile, there are several reports about the application in *Aspergillus oryzae* [54-56] and *Trichoderma reesei* [57] heterologous gene expression.

The Application in Mushrooms

Mushrooms, also known as the macrofungi, are another category of filamentous fungi. They can generate fruiting bodies and sclerotia. In fungal taxonomy, most of them belong to basidiomycotina. Common macrofungi include *Lentinus edodes*, *Volvariella volvacea*, *Agrocybe cylindracea*, *Auricularia auricula*, *Pleurotus ostreatus*, *Tremella fuciformis*, *Pleurotus eryngi*, *Hypsizigus marmoreus* etc. They are both important categories of fungi vegetables and important raw materials for food and pharmaceutical industries.

Large amount of transgenic research using mushrooms as hosts have been carried out for decades, especially in the field of transgenic technologies. So far, several transgenic methods for mushrooms have been established, such as $CaCl_2$-PEG-mediated genetic transformation [58], electroporation [59], gene gun transformation [60], restriction enzyme-mediated genetic transformation [61] and Agrobacterium-mediated genetic transformation [62, 63]. However, the expression levels of heterologous genes were low [63, 64]. Even though codon optimization has been widely used to improve the expression level of foreign genes in animals, plants and many other microorganisms and while many of these research studies have achieved highly expressed proteins, mushrooms have not been reported as a common expression system. Predictably, improving the genetic expression efficiency of exogenous genes by optimizing their codons with hosts' optimal codons will be one of the best strategies for high-level expression of heterologous genes in mushrooms in the future.

REFERENCES

[1] Crick FH, Barnett L, Brenner S (1961) General Nature of the Genetic Code for Proteins. *Nature* 192:1227-1232.

[2] Nirenberg MW, Matthaei JH (1961) The Dependence of Cell-free Protein Synthesis in *E.coli* upon Naturally Occurring or Synthetic Polyribonucleotides. *Proc. Natl. Acad. Sci. USA* 47:1588-1602.

[3] Kurland CG (1991) Codon Bias and Gene Expression. *FEBS* 285(2):165-169.

[4] Hershberg R, Petrov DA (2008) Selection on codon bias. *Annu. Rev. Genet.* 42:287-299.

[5] Jeffrey L, Bennetzen S, Hall BD (1982) Codon Selection in Yeast. *J. Biol. Chem* 257(6):3026-3031.

[6] Bulmer M (1991) The Selection-mutation-drift-theory of Synonymous Codon Usage. *Genetics* 12:897-907.

[7] Ikemura T (1981) Correlation between the abundance of *Escherichia coli* transfer RNAs and the occurrence of the respective codons in its protein genes. *J. Mol. BIOl* 146(1):1-21.

[8] Sharp PM, Tuohy TMF, Mosurski KR (1986) Codon usage in yeast: cluster analysis clearly differentiates highly and lowly expressed genes. *Nucleic. Acids Res.* 14(13):5125-5143.

[9] Sharp PM, Li WH (1987). The codon adaptation index—ameasure of directional synonymous codon usage bias, and its potential applications. *Nucleic. Acid Res.* 15(3):1281-1295.

[10] Ruiz LM, Armengol G, Habeych E, Orduz S (2006) A theoretical analysis of codon adaptation index of the *Boophilus microplus bm86* gene directed to the optimization of a DNA vaccine. *J. Theor. Biol.* 239(4):445-449.
[11] Liu QP (2006) Analysis of Codon Usage Pattern in The Radioresistant Bacterium *Deinococcus radiodurans. BioSystems* 85:99-106.
[12] Wright F (1990). The "efective number of codons" used in a gene. Gene, 87(1):23-29.
[13] Fuglsang A (2004). The "efective number of codons" revisited. *Biochem. Biophys. Res. Commun* 317(3):957-964.
[14] Wang K, Liu CQ, Cao H, Chen XF (2006) Insight into relationship among mRNA coding region length, folding tendency and codon usage bias in *Escherichia coli. Acta. Microbiol. Sinica.* 46(6):895-899.
[15] Zhao S, Zhang Q, Liu XL, Wang XM, Zhang HL, Wu Y, Jiang F (2008) Analysis of synonymous codon usage in 11 human bocavirus isolates. *Biosystems* 92:207-214.
[16] Warrick HM, Spudich JA (1998) Codon preference in *Dictyostelium discoideum. Nucleic. Acids Res.* 16(14A):6617-6635.
[17] Lavner Y, Kotlar D (2005) Codon bias as a factor in regulating expression via translation rate in the human genome. *Gene.* 345(1):127-138.
[18] Morton BR (1993) Chloroplast DNA codon use: evidence for se1ection at the *psb A* locus based on tRNA availability. *J. Mol. Evol.* 37(3):273-280.
[19] Lopez P, Philippe H (2001) Composition strand asymmetries in prokaryotic genomes: mutational bias and biased gene orientation.*C R Acad. Sci. III* 324(3):201-208.
[20] Hu GB, Zhang SL, Xu CJ, Lin SQ (2006) Analysis of codon usage between different citrus species. Journal of South China Agriculture University 27(1):13-16 (in Chinese with an English abstract).
[21] Zhou M, Tong CF, Shi JS (2007) Analysis of codon usage between different poplar species. *J. Genet. Genomics* 34(6):555-561.
[22] Fox JM, Erill I (2010) Relative codon adaptation: a generic codon bias index for prediction of gene expression. *DNA Res.* 17(3):185-196.
[23] Zhang Z, Li J, Cui P, Ding F, Li A, Townsend JP, Yu J (2012) Codon Deviation Coefficient: a novel measure for estimating codon usage bias and its statistical significance. *BMC Bioinformatics* 13(1):43-52.
[24] Wu XM, Wu SF, Ren DM, Zhu YP, He FC (2007). The analysis method and progress in the study of codon bias. Hereditas 29(4):420-426 (in Chinese with an English abstract).
[25] Zhao X, Li Z, Lu SF, Huo KK, Li YY (1999) Synonymous codon usage in *Yarrowia lipolytica.* Journal of Fudan University (Natural Science) 39(5):510-516(in Chinese with an English abstract).
[26] Carlini DB, Chen Y, Stephan W (2001) The relationship between third-codon position nucleotide content, codon bias, mRNA secondary structure and gene expression in the drosophilid alcohol dehydrogenase genes Adh and Adhr. *Genetics* 159(2):623-633.
[27] Stenstrom CM, Jin HN, Major LL (2001) Codon bias at the 3'-side of the initiation codon is correlated with translation initiation efficiency in *Escherichia coli. Gene.* 263(1-2):273-284.
[28] Karlin S, Campbell AM, Mrazek J (1998) Comparative DNA analysis across diverse genomes. *Annu. Rev. Genetics* 32:185-225.

[29] Shi X, Wang X, Li Z, Zhu Q, Tang W, Ge S, Luo J (2006) Nucleotide substitution pattern in rice paralogues: implication for negative correlation between the synonymous substitution rate and codon usage bias. *Gene*. 376(2):199-206.

[30] Duret L, Mouehireud D (1999) Expression pattern and, surprisingly, gene length shape codon usage in *Caenorhabditis, Drosophila,* and *Arabidopsis. Proc. Natl. Acad. Sci. USA* 96(8):4482-4487.

[31] Moriyama EN, Powell JR (1998) Gene length and codon usage bias in *Drosophila melanogaster, Saccharomyces cerevisiae* and *Escherichia coli. Nucleic. Acids Res.* 26(13):3188-3193.

[32] Eyre-Walker A (1996) Synonymous codon bias is related to gene length in Escherichia coli: selection for translational accuracy? *Mol. Biol. Evol.* 13(6):864-872.

[33] Marín A, González F, Gutiérrez G, Oliver JL (1998) Gene length and codon usage bias in *Drosophila melanogaster, Saccharomyces cervisiae* and *Escherichia coli. Nucleic. Acid Res.* 26(19):4540.

[34] Moriyama EN, Powell JR (1997) Codon usage bias and tRNA abundance in *Drosophila. J. Mot. Evol.* 45(5):514-523.

[35] Gupta SK, Bhattacharya TK (2000) Studies on the relationships between the synonymous codon usage and protein secondary structural units. *Biochem. Biophys. Res. Commun.* 269(3):692-696.

[36] Zama M (1990) Codon usage and secondary structure of mRNA. *Nucleic. Acids Symp.* Ser. 22:93-94.

[37] Toshimichi I (1985) Codon usage and tRNA content in unicellular and multicellular organisms. *Mol. Boil.* E 2(1):13 -34.

[38] Gao F, Li Y, Decker JM, Peyerl FW, Bibollet-Ruche F, Rodenburg CM, Chen Y, Shaw DR, Allen S, Musonda R, Shaw GM, Zajac AJ, Letvin N, Hahn BH (2003) Codon usage optimization of HIV type 1 subtype C *gag, pol, env,* and *nef* genes: in vitro expression and immune responses in DNA-vaccinated mice. *AIDS Res. Hum. Retroviruses* 19(9):817-823.

[39] Panahi M, Alli Z, Cheng X, Belbaraka L, Belgoudi J, Sardana R, Phipps J, Altosaar I (2004) Recombinant protein expression plasmids optimized for industrial *E.* coli fermentation and plant systems produce biologically active human insulin-like growth factor-1 in transgenic rice and tobacco plants. *Transgenic. Res.* 13(3):245-259.

[40] Dai Z, Chen Y, Qi W, Huang L, Zhang Y, Zhou T, Yang X, Gao G (2013) Codon optimization increases human kallistatin expression in *Escherichia coli. Prep. Biochem. Biotechnol.* 43(1):123-136.

[41] Wang Q, Mei C, Zhen HH and Zhu J (2012) Codon preference optimization increases prokaryotic cystatin C expression. J. Biomed. Biotechnol. doi:10.1155/2012/732017.

[42] Retnoningrum DS, Pramesti HT, Santika PY, Valerius O, Asjarie S, Suciati T (2012) Codon optimization for high level expression of human bone morphogenetic protein-2 in *Escherichia coli. Protein Expr. Putif.* 84(2):188-194.

[43] Boycheva SS, Bachvarov BI, Berzal-Heranz A, Ivanov IG (2004) Effect of 3' terminal codon pairs with different frequency of occurrence on the expression of cat gene in *Escherichia coli. Curr. Microbiol.* 48(2):97-101.

[44] Wu ZL, Qiao J, Zhang ZG, Guengerich FP, Liu Y, Pei XQ (2009) Enhanced bacterial expression of several mammalian cytochrome P450s by codon optimization and chaperone coexpression. *Biotechnol. Lett.* 31(10):1589-1593.

[45] Pan QH, Xie QL, Hong A, Pu HL, Liu GL, Liao MD (2002) Cloning and high expression of nattokinase with an effective strategy. *Microbiology China*, 2002, 29(6): 59-63.

[46] Wang X, Li X, Zhang Z, Shen X, Zhong F (2010) Codon optimization enhances secretory expression of *Pseudomonas aeruginosa* exotoxin A in *E. Coli*. *Protein Expr. Purif.* 72(1):101-106.

[47] Tang L, Jiang R, Zheng K, Zhu X (2011) Enhancing the recombinant protein expression of halohydrin dehalogenase HheA in *Escherichia coli* by applying a codon optimization strategy. *Enzyme. Microb. Technol.* 49(4):395-401.

[48] Hedfalk K (2012) Codon optimisation for heterologous gene expression in yeast. *Methods Mol. Biol.* 866:47-55.

[49] Oberg F, Ekvall M, Nyblom M, Backmark A, Neutze R, Hedfalk K (2009) Insight into factors directing high production of eukaryotic membrane proteins; production of 13 human AQPs in *Pichia pastoris*. *Mol. Membr. Biol.* 26:215–227.

[50] Yagi S, Yagi-Tanaka K, Yoshioka J, Suzuki M (1993) Expression enhancement of the Tn5 neomycin-resistance gene by removal of upstream ATG sequences and its use for probing heterologous upstream activating sequences in yeast.*Curr. Genet.* 24(1-2):12-20.

[51] Tu Y, Wang Y, Wang G, Wu J, Liu Y, Wang S, Jiang C, Cai X (2012) High-level expression and immunogenicity of a porcine circovirus type 2 capsid protein through codon optimization in *Pichia pastoris*. Appl. Microbiol. Biotechnol. doi:10.1007/s00253-012-4540-z.

[52] Gao Z, Li Z, Zhang Y, Huang H, Li M, Zhou L, Tang Y, Yao B, Zhang W (2012) High-level expression of the *Penicillium notatum* glucose oxidase gene in *Pichia pastoris* using codon optimization. *Biotechnol. Lett.* 34(3):507-514.

[53] Luang S, Hrmova M, Ketudat Cairns JR (2010) High-level expression of barley beta-D-glucan exohydrolase HvExoI from a codon-optimized cDNA in *Pichia pastoris*. *Protein Expr. Purif.* 73(1):90-98.

[54] Tanaka M, Tokuoka M, Shintani T, Gomi K (2012) Transcripts of a heterologous gene encoding mite allergen Der f 7 are stabilized by codon optimization in *Aspergillus oryzae*. *Appl. Microbiol. Biotechnol.* 96(5):1275-1282.

[55] Sasaguri S, Maruyama J, Moriya S, Kudo T, Kitamoto K, Arioka M (2008) Codon optimization prevents premature polyadenylation of heterologously-expressed cellulases from termite-gut symbionts in *Aspergillus oryzae*. *J. Gen. Appl. Microbiol.* 54(6):343-351.

[56] Tokuoka M, Tanaka M, Ono K, Takagi S, Shintani T, Gomi K (2008) Codon optimization increases steady-state mRNA levels in *Aspergillus oryzae* heterologous gene expression. *Appl. Environ. Microbiol.* 74(21):6538-6546.

[57] Te'o VS, Cziferszky AE, Bergquist PL, Nevalainen KM (2000) Codon optimization of xylanase gene *xynB* from the thermophilic bacterium *Dictyoglomus thermophilum* for expression in the filamentous fungus *Trichoderma reesei*. *FEMS Microbiol. Lett.* 190(1):13-19.

[58] Li G, Wang Q, Liu QY, Li BJ (2004) Establishment of a transformation system of *Ganoderma lucidum* using PEG method. *Mycosystema* 23(2):255-261.

[59] Guo L, Liu Y, Zhao S, Liu E, Lin J (2008) Highly efficient transformation of intact yeast-like conidium cells of *Tremella fuciformis* by electroporation. *Sci. China C Life Sci.* 51(10):932-940.

[60] Li A, Horgen P A (1993) Attempts to develop a transformation system in *Agaricus bisporus*, using particle bombardment and several other novel approaches. *Cult. Mush. Newsl.* 1:11-16.

[61] Xie BG, Lu QQ, Rao YB, Sun SJ, Zheng JG (2007) Synthesis of human insulin gene and transformation into *Tremella fuciformis*. *Acta. Edulis. Fungi.* 14(2):1-8.

[62] Guo YT (2010) Research on *Agrobacterium Tumefaciens*-mediated transformation of human insulin gene BCA to *Tremella Fuciformis*. The M.D. Dissertation of Fujian Agricultural and Forestory University.

[63] Zhang WR (2011) Different *Agrobacterium Tumefaciens*-mediated transformation of human insulin gene to *Tremella Fuciformis* and detecting expression products by ELISA. The M.D. Dissertation of Fujian Agricultural and Forestory University.

[64] Sun SJ, Chen DX, Xie BG, et al. (2009) Isolation of GPD promoter from *Tremella fuciformis* and driving expression of EGFP gene. *DNA and Cell Biology* 28(2): 65-70.

In: Biochemistry and Molecular Biology
Editor: Mary Miu Yee Waye
ISBN: 978-1-63482-312-8

Chapter 7

THE BLOOD-BRAIN BARRIER ROADBLOCK: STRATEGIES FOR TARGETED DRUG DELIVERY

Geoffrey H. Morris[1], Tessa N. Campbell[2] and Francis Y. M. Choy[1*]
[1]Centre for Biomedical Research, Department of Biology,
University of Victoria, Station CSC, Victoria, BC, Canada
[2]Faculty of Medicine, University of Calgary,
Calgary, Alberta, Canada

ABSTRACT

Despite ongoing medical advances, the development of successful treatments for neurological and neurodegenerative disease remains challenging. This is primarily due to the presence of the blood-brain barrier, a protective interface that reduces exposure of the brain to potential toxins. Unfortunately, this barrier acts as a roadblock for drug delivery, preventing transport of neurotherapeutics across the membrane to their targets within the brain. Numerous approaches have been undertaken to overcome the blood-brain barrier obstacle. Such strategies include: direct injection into the brain, therapeutic ultrasound, cell-penetrating peptides, and nanoparticles. In this review, we discuss the theory behind these approaches, as well as the advantages and limitations of each method.

Keywords: Blood-brain barrier, drug delivery, ultrasound, cell penetrating peptides, protein transduction domain, nanoparticles

INTRODUCTION

Brain parenchyma is separated from blood by the specialized blood-brain barrier (BBB), which exhibits brain-specific endothelial cell characteristics that restrict the passage of blood-borne molecules into the brain, but ensure the delivery of essential nutrients and selected

[*] Corresponding author: E-mail: fchoy@uvic.ca; Phone: 1-250-721-7107.

biomolecules. The principal features of the vertebrate BBB are closed cell-cell junctions, a low rate of transcytosis, and the expression of various solute carriers and adenosine triphosphate-binding cassette transporters (Keller 2013). Unfortunately, it is this protective combination of morphologic, physiologic, and functional characteristics that ultimately limits the entry of neurotherapeutics into the brain. Therefore, significant effort has been invested into the engineering of various techniques of bypassing or modifying the BBB to gain entry into the brain parenchyma.

In this review, we discuss the theory behind four main approaches to overcome the BBB roadblock: direct injection of drugs into the brain, therapeutic ultrasound to modify the BBB, and cell penetrating peptides and nanoparticles to facilitate entry through the BBB (Figure 1). Advantages and limitations of each method are also considered.

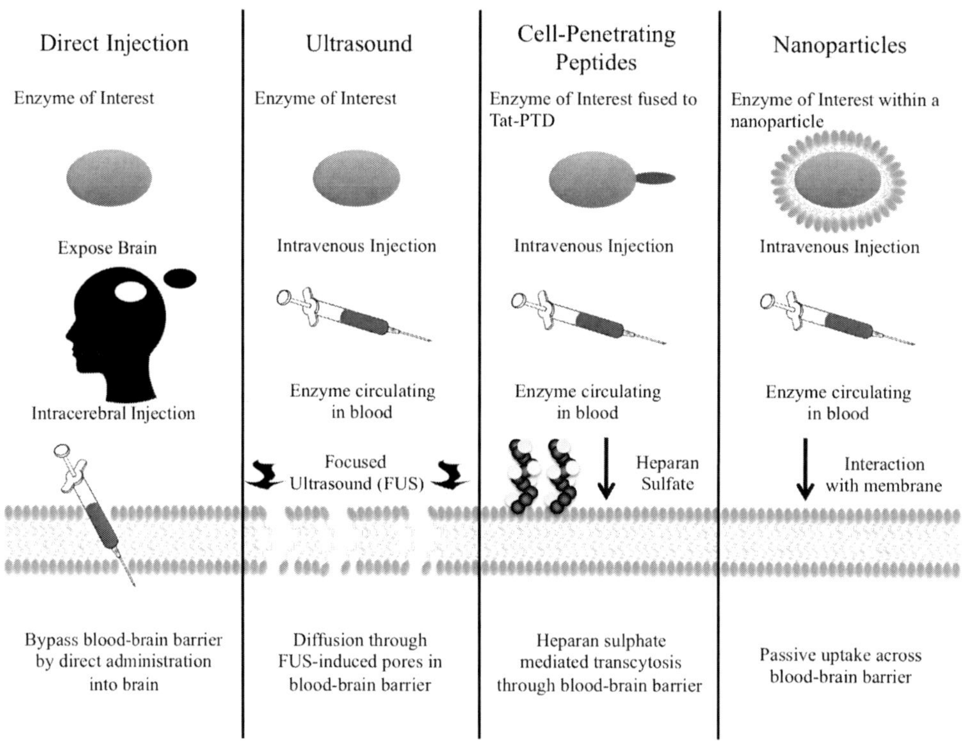

Figure 1. Representation of four methods used to deliver drugs across the blood-brain barrier.

Methods

Direct Injection

One approach to overcome the BBB roadblock is the administration of drugs directly into the brain. Direct delivery methods generally employ stereotactic positioning and specific procedures to locate target sites inside the brain in reference to anatomical landmarks, thus bypassing the BBB by injecting the treatment directly into the brain parenchyma (Deng

2010). The three most common approaches are: intracerebral implantation, intracerebroventricular infusion, and convection enhanced diffusion (Pardridge 2005).

Savas et al. (2004) explored this direct method of drug delivery for the treatment of Mucopolysaccharidosis (MPS) IIIA, an autosomal recessive disorder caused by a deficiency in the enzyme heparan N-sulfatase (sulfamidase). Using an MPS IIIA mouse model that was shown to have a disease progression similar to that of human patients, the authors delivered intracerebral injections of recombinant human sulfamidase directly into the murine dentate gyrus. This approach was found to delay neurodegenerative changes in young MPS IIIA mice (Savas et al. 2004). Further studies demonstrated that repeated doses of sulfamidase (Hemsley et al. 2007), or removal of mannose-6-phosphate residues from sulfamidase (Rozaklis et al. 2011) improved the effectiveness of treatment.

The main advantage of the direct injection technique is the bypassing of the BBB, thereby improving delivery and requiring less molecular and chemical modification of neurotherapeutics. However, disadvantages include limited therapeutic efficacy due to inadequate drug diffusion from the injection sites, potential damage to untargeted brain tissue from needle penetration, and risk of complications such as infection from invasive procedures (Deng 2010; Table 1).

Table 1. Advantages and disadvantages of four methods used to overcome drug delivery across the blood-brain barrier (BBB)

Method	Advantages	Disadvantages
Direct Injection	• Avoid difficulty of crossing the BBB	• Requires invasive surgery • Risk of infection and tissue damage
Ultrasound	• Can be targeted to any part of the body • Permeability of BBB is transient	• Mechanism of BBB permeabilization not fully understood • Potential tissue damage
Cell Penetrating Peptides	• Very effective in crossing the BBB • Can aid large molecules to cross the BBB	• Rapid clearance from the blood • Do not specifically target BBB endothelial cells
Nanoparticles	• Highly customizable • Can use ligands to target the BBB directly	• Unknown influences on toxicokinetics

Ultrasound

Ultrasound, defined as waves at a frequency at or above 20 kHz, can be focused, reflected, and refracted through different mediums (Kodama et al. 2006). Due to the unique characteristics of ultrasound, it is not surprising that this technique has been explored as a potential method of inducing BBB disruption for drug delivery into the brain (Deng 2010).

Two distinct ultrasound-based approaches have received the majority of attention. The first utilizes the thermal effects of focused ultrasound; the second employs mechanical effects and acoustic cavitation. Regarding the thermal approach, studies have demonstrated that the BBB was disrupted near brain tissue that had received high intensity focused ultrasound. This disruption was associated with neuronal damage and tissue necrosis (Vykhodtseva et al. 1995, Deng 2010).

The second approach utilizes the mechanical effects of acoustic cavitation initiated by preformed microbubbles in the brain vasculature. Effects such as shear stress due to bubble oscillation and shock waves from bubbles collapsing have been identified as important factors involved in increasing the permeability of cellular membranes and endothelial barriers (Postema et al. 2004, Rosenthal et al. 2004). At low acoustic power levels, the bubbles are under "stable oscillation" in which their size grows and diminishes (Choi et al. 2007). The result is small scale eddying that can affect only those endothelial cells that are near the active bubbles. At high acoustic power levels, the bubbles can grow rapidly and collapse violently through inertial cavitation. This level of activity can create extremely high pressure and temperature, high-velocity jets, and products of free-radicals (Deng et al. 2004). Through a combination of stable and inertial cavitation, many studies have found biological effects both *in vitro* and *in vivo* (Miller et al. 1996; Deng et al. 2004).

For example, Konishita et al. (2006) delivered a Herceptin monoclonal antibody to a targeted region of the mouse brain. Herceptin was the first antibody drug to be successfully used in the treatment of metastatic breast cancer; however, it had been unsuccessful in patients with brain tumours due to its inability to cross the BBB (Slamon et al. 2001). Using MRI-guided focused ultrasound BBB disruption, Herceptin was able to cross the barrier and reach the target tissue with minimal histological changes (Kinoshita et al. 2006).

Regarding the thermal approach, a window for BBB disruption without irreversible tissue damage has yet to be determined. Therefore, this method has considerable disadvantages compared to the mechanical approach. Conversely, the use of microbubbles offers a number of potential advantages: reduced ultrasound intensity resulting in less tissue damage, containment of disruptive cavitation events within the microvasculature, and less likelihood of irreversible neuronal damage (Deng 2010). However, the detailed mechanism for this second approach is yet to be elucidated, so further experiments are needed to determine its potential clinical success (Table 1).

Cell Penetrating Peptides

With the discovery of cell penetrating protein transduction domains (PTDs), new opportunities to develop therapeutic proteins fused to PTDs have surfaced. PTDs are small peptides capable of traversing the biological membrane and bringing their fusion partners along with them (Figure 1). Instead of injecting a virus that contains a gene of interest, the cell-penetrating portion is directly attached to a protein, avoiding the issue of viral integration. Several studies have shown the potential of PTDs in drug delivery (Jin *et al.* 2001), including the transduction of proteins as large as 110 kDa (Dietz & Bähr 2004).

One such example is the HIV Tat-PTD (Bandsmer *et al.* 2006; Jantzen *et al.* 2013). The functional 11 amino acid portion of Tat-PTD (YGRKKKRRQRRR) has been demonstrated by Schwarze et al., (1999) to transport fusion partners, including large proteins, across the BBB. In their study, recombinant protein containing the enzyme β-galactosidase fused to Tat-PTD was found to enter the mouse brain *in vivo* following intraperitoneal injection without any notable toxicity or immunogenicity from Tat-PTD. Moreover, an improved variant of Tat-PTD, named PTD4, has since been developed by Ho et al. (2001). PTD4 contains optimized α-helical structure and placement of basic arginine residues (amino acid sequence YARAAARQARA). This results in 33 times greater transduction *in vitro* and five times greater transduction *in vivo* over wildtype Tat-PTD (Ho *et al.* 2001).

Since the Schwarze et al. (1999) study, numerous other molecules have been shown to cross the BBB to positively affect the target site when fused to the HIV Tat-PTD (Zhang *et al.* 2012). For example, mice treated with a Neuroglobin-PTD fusion protein showed significantly less brain infarct volume and had better neurologic outcomes following middle cerebral artery occlusion when compared to a control group (Cai et al. 2011). Similar success has been shown with Bcl-x^L (Doeppner et al. 2010) and heat shock protein fusion products (Doeppner et al. 2009).

Though cell penetrating peptides have been successfully demonstrated to cross the BBB, they do not specifically target BBB cells. Moreover, they are rapidly cleared from the blood (Table 1). Thus, these specificity and pharmacokinetic properties must be taken into consideration during therapeutic design and development stages (Munyendo et al. 2012).

Nanoparticles

In the past several years, there have been a growing number of publications on nanotechnology and its use in drug delivery. Nanotechnologies are devices and materials that have a functional organization in at least one dimension on the nanometer scale (Silva 2008). It has been demonstrated that nanoparticles can be used to maintain drug levels in a therapeutically desirable range, and increase half-lives, solubility, stability, and permeability of drugs. Nanoparticles can be structurally adapted to deliver a variety of drugs, improve delivery efficiency, and reduce side effects by targeted delivery (Yang 2010).

Several major types of nanoparticles widely used for construction of nanomedicines, such as liposomes, dendrimers, and polymeric micelles, have also found successful applications in the central nervous system (Yang 2010). Therefore, the potential biomedical and pharmacological applications of nanoparticles for delivery of drugs across the BBB have received considerable attention.

For example, Afergan et al. (2008) described an innovative use of liposomes as a secondary vehicle for delivery of serotonin. Serotonin-encapsulated liposomes were phagocytosed by monocytes. With the permeation of monocytes across the BBB, phagocytosed serotonin was selectively delivered to the brain. In another example, the brain-targeting ligand lactoferrin was coupled to a dendrimer for directed delivery of DNAs. The lactoferrin-coupled dendrimer was reported to have high BBB-crossing efficiency compared to unmodified counterparts (Huang et al. 2009).

Despite their high specificity and multifunctionality, nanomedicines should be further scrutinized prior to widespread clinical application. Safety concerns regarding acute or long-term toxic effects must be thoroughly addressed (Table 1).

Conclusion

The blood-brain barrier, a protective interface that reduces exposure of the brain to potential toxins, serves as a roadblock that restricts the entry of neurotherapeutics. Numerous approaches have been undertaken to overcome or bypass this blood-brain barrier obstacle. Such strategies include: direct injection of drugs into the brain, therapeutic ultrasound to modify the BBB, and cell penetrating peptides and nanoparticles to facilitate entry through the BBB. Each approach has its own advantages and limitations, therefore a complementary combination of technologies may be needed to deliver the drug and carry out other key functions, including shielding the active drug from producing systemic side effects, avoiding premature clearance and metabolism, and targeting specific cells after entry into the brain. This multifaceted strategy may serve to significantly improve the quality of life for those suffering from neurodegenerative and neurological disorders, and reduce the burden of disease on society.

Acknowledgment

This work was supported by a research operating grant from the Sanfilippo Children Research Foundation to F.Y.M. Choy, and a graduate research fellowship from the University of Victoria to G. H. Morris.

References

Afergan, E., Epstein, H., Dahan, R., Koroukhov, N., Rohekar, K., Danenberg, H.D. & Golomb, G. (2008). Delivery of serotonin to the brain by monocytes following phagocytosis of liposomes. *J. Control Release*, 132, 84–90.

Bandsmer, J.C., Campbell, T.N., Cheyne, I. & Choy, F.Y.M. (2006). Expression of active α-*N*-acetylglucosaminidase/TAT chimerae in cultured *Spodoptera frugiperda* cells. *Protein and Peptide Letters*, 13, 353-356.

Cai, B., Lin, Y., Xue, X.-H., Fang, L., Wang, N. & Wu, Z.-Y. (2011). TAT-mediated delivery of neuroglobin protects against focal cerebral ischemia in mice. *Experimental Neurology*, 227, 224–231.

Choi, J.J., Pernot, M., Small, S.A. & Konofagou, E.E. (2007). Noninvasive, transcranial and localized opening of the blood-brain barrier using focused ultrasound in mice. *Ultrasound in Medicine & Biology*, 33, 95–104.

Deng, C.X., Sieling, F., Pan, H. & Cui, J.M. (2004). Ultrasound-induced cell membrane porosity. *Ultrasound in Medicine & Biology*, 30, 519–526.

Deng, C.X. (2010). Targeted drug delivery across the blood-brain barrier using ultrasound technique. *Ther Deliv.*, 1(6), 819-848.

Dietz, G.P.H. & Bähr, M.M. (2004). Delivery of bioactive molecules into the cell: the trojan horse approach. *Molecular and Cellular Neuroscience*, 27, 85–131.

Doeppner, T.R., Dietz, G.P.H., Weise, J. & Bähr, M.M. (2010). Protection of hippocampal neurogenesis by TAT-Bcl-X_L after cerebral ischemia in mice. *Experimental Neurology*, 223, 548–556.

Doeppner, T.R., Nagel, F., Dietz, G.P.H., Weise, J., Tönges, L., Schwarting, S. & Bähr, M.M. (2009). TAT-Hsp70-mediated neuroprotection and increased survival of neuronal precursor cells after focal cerebral ischemia in mice. *Journal of Cerebral Blood Flow and Metabolism*, 29, 1187–1196.

Hemsley, K.M., King, B. & Hopwood, J.J. (2007). Injection of recombinant human sulfamidase into the CSF via the cerebellomedullary cistern in MPS IIIA mice. *Molecular Genetics and Metabolism*, 90, 313–328.

Ho, A., Schwarze, S.R., Mermelstein, S.J., Waksman, G. & Dowdy, S.F. (2001). Synthetic protein transduction domains: enhanced transduction potential in vitro and in vivo. *Cancer Research*, 61, 474–477.

Huang, R., Ke, W., Han, L., Liu, Y., Shao, K., Ye, L., Lou, J., Jiang, C. & Pei, Y. (2009). Brain-targeting mechanisms of lactoferrin-modified DNA-loaded nanoparticles. *J Cereb Blood Flow Metab.*, 29,1914–23.

Jantzen, R.R., Truelson, S.N., & Choy, F.Y.M. 2013. Human α-N-acetylglucosaminidase: cDNA cryptic site removal and native secretion signal addition significantly enhance enzyme expression and secretion. *Process Biochemistry* 48:1197-1202.

Jin, L.H., Bahn, J.H., Eum, W.S., Kwon, H.Y., Jang, S.H., Han, K.H., Kang, T.C., Won, M.H., Kang, J.H., Cho, S.W., Park, J.K. & Choi, S.Y. (2001). Transduction of human catalase mediated by an HIV-1 TAT protein basic domain and arginine-rich peptides into mammalian cells. *Free Radical Biology and Medicine*, 31, 1509–1519.

Keller, A. (2013). Breaking and building the wall: the biology of the blood-brain barrier in health and disease. *Swiss Med Wkly*, 143, w13892.

Kinoshita, M., McDannold, N., Jolesz, F.A. & Hynynen, K. (2006). Noninvasive localized delivery of herceptin to the mouse brain by MRI-guided focused ultrasound-induced blood–brain barrier disruption. *Proceedings of the National Academy of Sciences*, 103, 11719–11723.

Kodama, T., Tomita, Y., Koshiyama, K.-I. & Blomley, M.J.K. (2006). Transfection effect of microbubbles on cells in superposed ultrasound waves and behavior of cavitation bubble. *Ultrasound in Medicine & Biology*, 32, 905–914.

Miller, M.W., Miller, D.L. & Brayman, A.A. (1996). A review of in vitro bioeffects of inertial ultrasonic cavitation from a mechanistic perspective. *Ultrasound in Medicine & Biology*, 22, 1131–1154.

Munyendo, W.L.L., Lv, H., Benza-Ingoula, H., Baraza, L.D. & Zhou, J. (2012). Cell penetrating peptides in the delivery of pharmaceuticals. *Biomolecules*, 2, 187-202.

Pardridge, W.M. (2005). The blood-brain barrier: bottleneck in brain drug development. *NeuroRx*, 2, 3–14.

Postema, M.M., van Wamel, A.A., Lancée, C.T.C. & de Jong, N.N. (2004). Ultrasound-induced encapsulated microbubble phenomena. *Ultrasound in Medicine & Biology*, 30, 14–14.

Rosenthal, I.I., Sostaric, J.Z.J. & Riesz, P.P. (2004). Sonodynamic therapy–a review of the synergistic effects of drugs and ultrasound. *Ultrasonics Sonochemistry*, 11, 349–363.

Rozaklis, T., Beard, H., Hassiotis, S., Garcia, A.R., Tonini, M., Luck, A., Pan, J., Lamsa, J.C., Hopwood, J.J. & Hemsley, K.M. (2011). Impact of high-dose, chemically modified sulfamidase on pathology in a murine model of MPS IIIA. *Experimental Neurology*, 230, 123–130.

Savas, P.S., Hemsley, K.M. & Hopwood, J.J. (2004). Intracerebral injection of sulfamidase delays neuropathology in murine MPS-IIIA. *Molecular Genetics and Metabolism*, 82, 13–13.

Schwarze, S.R., Ho, A., Vocero-Akbani, A. & Dowdy, S.F. (1999). In vivo protein transduction: delivery of a biologically active protein into the mouse. *Science*, 285, 1569–1572.

Silva, G.A. (2008). Nanotechnology approaches to crossing the blood-brain barrier and drug delivery to the CNS. *BMC Neuroscience*, 9 (Suppl 3), S4.

Slamon, D.J., Leyland-Jones, B., Shak, S., Fuchs, H., Paton, V., Bajamonde, A., Fleming, T., Eiermann, W., Wolter, J., Pegram, M., Baselga, J. & Norton, L. (2001). Use of chemotherapy plus a monoclonal antibody against HER2 for metastatic breast cancer that overexpresses HER2. *New England Journal of Medicine*, 344, 783–792.

Vykhodtseva, N.I., Hynynen, K. & Damianou, C. (1995). Histologic effects of high intensity pulsed ultrasound exposure with subharmonic emission in rabbit brain *in vivo. Ultrasound Med Biol.*, 21(7), 969-979.

Yang, H. (2010). Nanoparticle-medicated brain-specific drug delivery, imaging and diagnosis. *Pharm Res.*, 27(9), 1759-1771.

Zhang, X., Zhang, X. & Wang, F. (2012). Intracellular transduction and potential of Tat PTD and its analogs: from basic drug delivery mechanism to application. *Expert Opinion on Drug Delivery*, 9, 457–472.

In: Biochemistry and Molecular Biology
Editor: Mary Miu Yee Waye
ISBN: 978-1-63482-312-8

Chapter 8

GENETICS AND EPIGENETICS OF BIPOLAR DISORDER

Ming Wai Yeung*[*1] and Miu Yee Mary Waye[2]
[1]School of Life Sciences
[2]School of Biomedical Sciences, Croucher Laboratory for Human Genomics,
The Chinese University of Hong Kong, Shatin, N.T. Hong Kong

ABSTRACT

Bipolar disorder (BD), also known as manic depression, is a severe mood disorder that causes shifts in mood, energy and ability to function. While there is no cure currently, many people with bipolar disorder could be treated and with the right medication, could go into remission and be symptom free for years or decades. It is estimated that BD affects 2.4% of the population worldwide. Patients of BD suffer from recurrent episodes of mania and depression with each episode typically lasting for 1 to 2 weeks. As a result, patients' lives are severely disrupted. Studies have shown that it is a complex mental disease which is highly heritable, polygenic yet multifactorial. So far vast numbers of candidate genes have been implicated for BD and several models have been proposed to explain the pathology of BD: ion channel pathology (ANK3, CACNA1C and GABRB2), disruption of signaling pathways (BDNF, DGKH, NTRK2 and NRG1), abnormal cell-cell adhesion (FAT1, CTNNA2, CDH7 and NCAN), and disruption in circadian rhythm (CLOCK, ARNNTL and TIMELESS). Concerning the epigenetic studies of BD, differential DNA methylation has been observed in both coding genes (HTR1A, MB-COMT CTNNA2 and MHR1) and non-protein coding gene (HCG9). In addition to specific gene candidates, some studies attempt to find the link between BD and epigenetic modification.

Keywords: GWAS, SNPs, Association study, DNA methylation, ion channel, signaling pathway, cell adhesion

[*] Corresponding Author: Ming Wai YEUNG, School of Life Sciences, The Chinese University of Hong Kong. Shatin, N.T., Hong Kong. Email: mwyeung@cuhk.edu.hk

LIST OF ABBREVIATIONS

ADHD: Attention deficit-hyperactivity disorder;
ASD: Autism spectrum;
BD: Bipolar disorder;
BPI: Bipolar I disorder;
BPII: Bipolar II disorder;
GWAS: Genome-wide association study;
MDD: Major depressive disorder;
SCZ: schizophrenia

INTRODUCTION

Major Symptoms of Bipolar Disorder

Bipolar disorder (BD), formerly called Manic-depressive Disorder, is a severe mental disorder in which patients suffer from recurrent episodes of mania and depression. People diagnosed as BD suffer from at least one onset of manic episode and one onset of a depressive episode. Symptoms of a manic episode include clearly elevated or irritable mood, restlessness, pressured speech, impulsive behavior and reduced need of sleep. In contrast, in depressive episodes patients have low or irritable moods. They lack energy and often show hypersomnia. They also lose interest in life or even attempt to commit suicides.

The severity as well as the time-length of each episode of mania or depression may vary. Each episode can last continuously from several days up to several months while in between the episodes patients behave normally. As a result, patients' lives are usually greatly disrupted by such mood shifts. BD has currently no cure and hence many patients of BD often rely on life-long medications to control their conditions.

Subtypes of Bipolar Disorders

Diagnostic and Statistical Manual of Mental Disorders, Fourth Edition (DSM IV), published by the American Psychiatric Association, classified BD into four subtypes:

Bipolar I Disorder (BPI), Bipolar II Disorder (BPII), Cyclothymic Disorder and Bipolar disorder Not Otherwise Specified (BP-NOS). Bipolar I is characterized by dominating manic episodes of more than one week or with severe manic symptoms. Patients of BPI also have depressive episodes. Bipolar II is different from Bipolar I in that BPII is characterized by less severe manic episodes. Patients of BPII generally have mood shift from hypomania to depression. Cyclothymic Disorder is a milder form of Bipolar disorder. Patients are diagnosed of Cyclothymic Disorder when their moods swing between episodes of mild mania and mild depression for at least two years and during this course absence of symptoms never exceeds two months for adult patients. The last subtype is Bipolar disorder Not Otherwise Specified (BP-NOS). BP-NOS patients demonstrate milder symptoms than those of BPI and BPII that they do not meet the diagnostic criteria (as in DSM IV) for either.

Prevalence and Age of Onset

BD is a common mental disorder. The World Health Organization World Mental Health survey initiative estimated BD affects 2.4% of the population worldwide on average [1]. Onset of BD usually happens in adolescence or early adulthood, with more than half of the patients experiencing their first onset before the age of 25. The mean age of onset is 18 for BPI, 20 for BP-II and 22 for milder forms of BP.

Personal and Social-Economic Impacts

Not only does bipolar disorder (BD) greatly affect the patient, it also causes a burden on the society. On a personal level, BD greatly disrupts the lives of patients given the rather extreme symptoms and its life-long recurrence. Patients usually face challenges in work, inter-personal relationship as well as in their financial responsibilities. BD significantly reduces patients' work productivity. For employed BP patients, they missed 11.5 extra workdays each year on average, when compared to their counterparts without serious psychiatric disorders [2]. On another study investigating disability in mental diseases [3], authors reported that BD patients have 20.4 additional days of partial disability (partial impairment in daily function) per year, which is higher than those suffering from other common diseases like major depression, cardiovascular disease or respiratory disease. An Australian study published in 2012 [4] estimated that the total lifetime cost associated to BD to be AUD$160,671 for BPI and AUD$94,401 for BPII; as compared with AUD$95,584 for unipolar disorder. BD also poses quite a substantial economic burden on the society. This is attributed by direct medical costs such as expenses for hospitalization, or indirect cost such as manpower loss, demand of caretakers and cost on suicidal attempts. Unemployment rate is usually higher among individuals with BD and some of them receive disability payments [5]. The annual cost of BD was estimated to be 342 million in UK [6]. A study in 1998 estimated BD costs US$24 billion in USA in the year 1998 alone [7] though another study proposed a much higher annual cost of BD – US$151.0 billion [8]. Nevertheless, these studies all pointed to a high socio-economic cost of BD to the society, even though the prevalence of BD may not be as high as common diseases like atherosclerosis or different forms of cancer.

High Heritability of Bipolar Disorder

Similar to other psychiatric disorders such as schizophrenia and autism, Bipolar disorder (BD) is a very complex disease. Amongst the various etiology of BD, heritable elements play an important role in BD. Population-based studies have estimated heritability of BD to be 0.59 to 0.62 ((9) (10) while twins studies suggested a higher value 0.71-0.93. [11, 12, 13]. In a nationwide study of probands and their offspring in Denmark [14], it was found that offspring with both parents diagnosed with BD had an approximately 51.9-fold higher risk in developing BD (24.9%) compared with the offspring with none of parents ever diagnosed with BD (0.48%) by the age of 52. Offspring with either parent having BD also had a 9.2-fold risk (4.4%) compared to those who have no BD-affected parents. Such large differences in risk of different groups of offspring suggest that the genetic makeup contribute to BD. On the other hand, twin studies [11, 12, 13] showed that monozygotic twins, who share the same genetic make-ups, did not always suffer BD as their bipolar twin siblings. This suggests that the genetic factor is one important factor, but it is not the only factor affecting the etiology of BD. Given that there is no one specific gene identified to be solely responsible for BD, BD shall be the effect of multiple genes as well as other factors like environmental factors. Nonetheless, understanding the genetic and epigenetic links to BD would help us understand the etiology of BD. This article will summarize the recent findings on the genetics and epigenetics study of BD and discuss some possible mechanism on the pathology of BD.

APPROACHES IN SEARCHING FOR CANDIDATE GENES RELATED TO SUSCEPTIBILITY OF BIPOLAR DISORDER

Family-Based Studies and Twin Studies

It is not an easy task to identify the genetic links to a certain disease, considering there is on average about 4 million genetic variants in each individual's genome [15]. Therefore, to study people with close biological relationship who share more similar genetic makeup would narrow down the candidate gene pool and ease the search of diseased-related genes. Family studies and twin studies are common and useful approaches, by which many research scientists have already reported the high heritability of BD.

Family Studies

The principle of family studies is that, for a genetic disease the culprit gene(s) pass through generations, a familial aggregation of the disease is often seen. By comparing the genomes of the diseased family members and healthy family members, there will be a higher chance to find the gene(s) associated than that obtained by comparing the affected with random controls. This approach has been used in linkage studies as well as the association of specific genes and yielded useful results. However, there are also shortcomings for this approach. One difficulty for researchers is to recruit enough numbers of family for research so the sample sizes are usually modest, which definitely would affect the significance of the findings. Another problem is the failure to replicate the results in different studies due to the genetic heterogeneity and polygenic nature of BD. The gene detected in one study might be a more important risk factor in a certain population, but not in the others and thus generating inconsistent results in different studies. To overcome these difficulties, meta-analysis seems to be a suitable solution. It combines the data from multiple studies to increase the sample size and hence generate a more convincing result. For instance, a meta-analysis of 11 linkage studies reported linkage of BPI to chromosome 6q and BPII to chromosome 8q with genome-wide significance (LOD>3.03). [16]. This linkage meta-analysis has provided directions for the search of candidate genes of BD and several follow-up association studies have since been carried out to investigate the specific genes on those regions [17, 18].

Twin Studies

Monozygotic twins are valuable sources for the studies of disease pathologies. Concordance studies on monozygotic twins and dizygotic twins, for instance, allowed researchers to determine genetic influences in a particular disease. In complex diseases like BD, however, it is the discordance between the monozygotic twins that is becoming an increasing interest topic for studies. Discordance between monozygotic twins may arise from rare but possible genetic changes including *de novo* point mutations, insertion/deletion, as well as epigenetic differences like differential DNA methylations, histone medication and other mechanisms [19]. In particular, monozygotic twins are ideal subjects for epigenetic studies because it minimizes the effects from genetic differences [20]. However the main problem in discordant twins study is the limited sample size. This may cause difficulties in producing statistically significant results. One may also question whether the results found specifically in some twins still hold true for the whole population. Hence, twin studies may be a powerful approach to identify the candidate genes but follow-up studies on a larger scale are needed to validate the findings.

Case-Control Studies

In case control studies subjects, biological relationship is not important. Subjects are recruited in a case group if they are affected by BD while individuals without mental diseases are recruited as controls. The sample size of a case-control study is usually bigger than that of a family study and a twin study. The case-control approach can be applied on gene association studies, genome wide association study (GWAS) as well as gene expression analyses.

Gene Association Studies

The aim of gene association studies is to investigate if a specific allele of a certain gene is related to the disease which a higher allele frequency is observed in the case group than that of the control group. Based on the evidence from previous studies or researchers' hypotheses, specific genetic markers are chosen to conduct the studies.

Genome Wide Association Studies (GWAS)

The basic principle of GWAS is that if there is a genetic influence in BD, there should be a common allele – a marker that is more frequent among the affected individuals than that of the healthy controls [21]. Like association studies of specific genes, the aim of GWAS is to link genetic variation to the disease of interest. But the difference of GWAS is that it does not have a hypothesis or a particular gene candidate for the study. Instead, a large number of genetic markers – SNPs (in the scale of million) which are chosen to represent the entire genome are investigated in a large number of subjects in one study. An advantage of GWAS is that it eliminates the bias of the researchers and thus sometimes leads to the discovery of new candidates that are beyond the current knowledge of BD [22]. However; there are also limitations for GWAS. First, GWAS usually rely on the use of microarrays which tested a fixed set of SNPs. This means rare variants that are not included in the array may be missed.

Second, GWAS alone cannot explain the link of genetic markers or the gene variants to the disease. Further investigations must be done to understand biological effects of those genes.

Gene Expression Studies

Gene expression studies investigate the expression level of a specific gene or a whole array of genes, i.e., mRNA level and protein level in a particular tissue, preferably in the brain for the study of BD. Changes of mRNA level and protein levels, unlike the DNA variation, may be dynamic, i.e., it may vary in different tissues, different time and it could be affected by external influences such as medications. Nevertheless, expression studies can provide clues to understanding the pathology of BD and possible drug interactions. It can also complement the results from gene association studies. For instance, both a gene association study and a gene expression study have implicated SLC1A1, gene that codes for neuronal glutamate transporter 1, in BD [23, 24].

Use of Animal Models

Animal models are useful for the study of numerous diseases including BD. There have been various genetically engineered mouse models to study the genetic basis of BD. For example, one study found that DGK-beta knockout mice exhibited hyperactivity and the effect can be reversed by lithium administration. Another study using mice found that reduced ankyrin G (encoded by gene ANK3) expression caused manic-like behavior [26]. ANK3 gene has been implicated in BD in numerous GWAS [22, 27, 28, 29] and this animal study may shed light on how ANK3 may participate in BD pathology with mood shift as a major characteristic.

Findings in Genetic Studies

Integrating the Findings into Current Models of BD Pathology

There have been increasing numbers of studies trying to identify risk genes for BD [30]. With the introduction of GWAS in the study of BD during the last decade, over ten GWAS have been carried out [31]. Owing to the hypothesis-free nature of GWAS, many new candidate genes which were rarely the targets of traditional gene association studies have been discovered. GWAS has most definitely expanded our knowledge of BD. However, as mentioned before, GWAS "tags" the gene only and it does not explain the link of the gene variants to the disease.

Table 1. Candidate genes in bipolar disorder studies

Genetic variants	Studies supporting association				Studies against association	
	GWAS	p-value	Association study	p-value	GWAS	Association
ANK3						
rs10994336	Ferreira,2008 Scott ,2009 Schulze,2009	9.1×10^{-9} 0.042 1.7×10^{-5}	Lett, 2011	0.046	Green, 2012 Takata,2011	Gella,2011
rs1938526	Ferreira,2008 Lee,2011 Takata,2011	1.3×10^{-8} 6.55×10^{-5} 0.048				
rs9804190	Schulze,2009	3×10^{-6}				Gella, 2011
rs10761482			Gella, 2011	0.015		
CACNA1C						
rs72552065					Sklar, 2002	
rs1006737	Ferreira,2008 Green, 2012	7.0×10^{-8} 4.09×10^{-4}	Lett, 2011	0.017	Sklar, 2008	
rs1024582	Ferreira,2008 Cross-Disorder Group of the Psychiatric Genomics Consortium, 2013	1.7×10^{-7} 1.87×10^{-8}				
rs4765913	Psychiatric GWAS Consortium Bipolar Disorder Working Group. 2011	1.52×10^{-9}				
rs4765905	Hamshere,2012	1.23×10^{-8}				
CACNB2						
rs2799573	Cross-Disorder Group of the Psychiatric Genomics Consortium, 2013	4.29×10^{-8}				
rs11013860	Lee, 2011	5.15×10^{-5}				
CACNB3						
rs11168751	Psychiatric GWAS Consortium Bipolar Disorder Working Group. 2011	2.51×10^{-6}				
CACNA1D						
rs3774609	Psychiatric GWAS Consortium Bipolar Disorder Working Group. 2011	3.73×10^{-5}				

Table 1. (Continued)

Genetic variants	Studies supporting association				Studies against association	
	GWAS	p-value	Association study	p-value	GWAS	Association
GABRB1						
rs7680321	The Wellcome Trust Case Control Consortium,2007	$6.2x10^{-5}$				
GABRB2						
rs3850733			Perlis,2008	<0.005		
KCTD12						
rs2073831	Lee,2011	$9.74 \times 10^{-6)}$				
BDNF						
rs6265			Neves-Pereira, 2002 Sklar, 2002 Green, 2006 Lee,2013 Wang, 2012*	0.00064 0.066# 0.004 0.007 0.025		
NTRK2						
rs2769605	Smith,2009	4.5×10^{-5}				
DGKH						
rs1012053	Baum, 2008 Yosifova, 2011	$1.5x 10^{-8}$ 4.50×10^{-2}				Tesli,2009
rs9315885			Ollila,2009 Squassina,2009	0.069# 0.073#		
rs2122246			Zeng, 2011	0.0186		
rs1170169			Weber, 2011	0.013		
rs9525580			Weber, 2011	0.007		
rs9315897			Takata, 2011	0.039		
NRG1			Green,2005 Prata,2009 Goes,2009			Cassidy,2006 Crisafulli,2012
FAT1						
rs455219	Belmonte Mahon,2011	$1.5x10^{-4}$				

Genetic variants	Studies supporting association				Studies against association	
	GWAS	p-value	Association study	p-value	GWAS	Association
rs2637777			Abou Jamra R, 2008	2.8×10^{-4}		
rs2304865			Blair, 2006	2×10^{-4}		
rs1298865			Blair, 2006	3×10^{-3}		
CTNNA2						
rs13409348	Scott,2009	2.7×10^{-6}				
CDH7						
rs1444067	Sklar, 2008/	7.26×10^{-5}	Soronen,2010	0.018		
rs2850700		6.02×10^{-5}		0.012		
rs2658046		5.34×10^{-5}		0.021		
rs12970791		5.55×10^{-5}		0.044		
rs2850699		6.02×10^{-5}		0.019		
rs4455070		4.10×10^{-5}				
NCAN						
rs1064395	Cichon,2011	3.02×10^{-8}				

#p-value after correction, which did not reach statistical significance ($p<0.05$)

*The risk allele proposed by this team was different from the others listed in the table

One may, therefore, question how these new gene candidates, implicated in GWAS, can be incorporated with other studies to enhance our understanding of BD. In the following sessions, emergent genes of interest in the study of BD which appear in the last decade will be discussed. These genes are involved in several models of BD pathology - ion channelopathy, signaling pathway, cell adhesion problem and circadian rhythm. The major genes candidates supported by GWAS, as well as other studies, will be discussed under these four models, but it should be noted that these networks are not completely independent of the others and there are clearly interactions between them. Such classification is for the ease of discussion.

A summary of BD-associated genes discussed here is listed in Table 1.

GWAS Results Point to Ion Channelopathy: ANK3 and CACNA1C

It is already known that ion channels are involved in several neurological diseases such as epilepsy and episodic ataxia [32] but most are muscle disorders and the roles of different ion channels in muscle contraction are well studied. Neuropsychiatric diseases like schizophrenia, autism and bipolar disorder (BD), however, have much more complicated phenotypes and the etiologies remain largely unknown. Gargus proposed in 2006 that genes related to ion channel functions may also play a role in neuropsychiatric disease [33]. The idea of ion channelopathy in BD was supported by a GWAS which found a strong association in ANK3 and CACNA1C with BD in 2008 [29]. ANK 3 codes for ankyrin G – an adaptor protein found on membranes of neurons whereas CACNA1C codes for a subunit of L-type voltage-dependent calcium channel.

ANK3

Ferreira and colleagues identified ANK3 (rs10994336, p= $9.1x10^{-9;}$ rs1938526 p= $1.3x10^{-8}$) and CACNA1C (rs1006737, p= $7.0x10^{-8}$; rs1024582, p= $1.7x10^{-7}$) as susceptibility genes for BD in UK and US populations [29]. Associations of the ANK3 variants in Ferreira's study reached threshold for genome-wide significance ($p \leq 5 \times 10^{-8}$). Subsequent GWAS generally supported the association of ANK3 with BD. Association of SNP rs10994336 was replicated by Scott et al. using different samples from Europe (p= 0.042) [34]. Schulze and colleagues replicated the association of SNP rs10994336 with German and US samples; they also reported another variants of ANK3 (rs9804190, p= $3x10^{-6}$) to be linked to BD [28]. However, a recent GWAS done by Green and colleagues using a new set of BD patients from UK failed to replicate the association of SNP rs10994336 (p= 0.912) [22].

Concerning the association of ANK3 to BD in other populations, Lee and colleagues in their GWAS with 1,000 Han BD patients from Taiwan reported association of one SNP near the SNP rs1938526 of ANK3 gene (p=6.55×10^{-5}) in Chinese Han population [35]. Takata and colleagues later carried out a GWAS combining the samples from Taiwan, Korea and Japan, they found an association of rs1938526 (p= 0.048) but not rs10994336 (p= 0.591) to BD [27]. Nevertheless the study showed ANK3 is associated with BD in East Asian population, but the genetic risk variant may be different from that of the western counterpart.

Following the GWAS in 2008, ANK3 became a popular target in the genetic study of BD. A family-based association study in Toronto found a marginal association of rs10994336 variant with non-psychotic BD (p=0.046) but the association was lost when psychotic BD patients were included [36]. At the same year, Gella and colleagues performed a case-control gene association study with German bipolar patients [37]. They found no association of rs10994336 and rs9804190 with BD either. However, they found an association of another ANK3 variant - rs10761482 with bipolar disorder (p=0.015), this variant was previously reported to associate with schizophrenia in a GWAS (38). Both association studies found no direct relationships of the common variants (rs10994336, rs9804190 and rs10761482) to BD. But it should be pointed out that considering the complexity of BD (polygenic and multifactorial), the effect contributed by ANK3 variants alone towards BD might be small and it might be masked by epistasis as well and thus leading to discrepancies in different studies.

In addition to association studies, two phenotypic studies implicated the risk allele of ANK3 (rs10994336 allele "T") in altered signal detection [39] and stress signal processing [40]. An expression study found a correlation between another common variant rs1938526 and altered ANK3 expression in brain [41]. The study found that there was a reduced transcript level for a brain specific Ankyrin G isoform in the postmortem brain tissues of individuals carrying minor allele "C" at SNP rs1938526. In a recent animal study, expression of ANK3 in brain of mice was reduced by RNA interference to investigate how ANK3 expression might affect mice behavior [26]. It was found that mice with reduced ANK3 expression demonstrated elevated activity and this change was reversed by lithium treatment. The same study also observed that chronic stress caused ANK3+/– mice to shift from more excited state to a depressive state while this shift did not occur in ANK3 +/+ mice. Mood shift is one prominent characteristic of BD when compared with other psychiatric diseases and there is little understanding in the molecular basis that causes the shift. This mouse study therefore demonstrated a possible biochemical link involving mood changes in response to environmental stimuli.

CACNA1C and Other Channel-Related Genes

Another popular candidate is CACNA1C which codes for alpha 1C subunit of a voltage-dependent calcium channel (Ca_v 1.2). The channel is expressed in muscles and brain [42]. It is thought the channel is important for regulation of neuronal activities by controlling the influx of calcium ions. Initially, an association study in 2002 was unable to find the association of CACNA1C with BP using the SNP rs72552065, which lies on exon 44 of the gene [43]. However following the emergence of GWAS, CACNA1C has once again become an interest in the study of BD. In 2008, Ferreira et al. reported 2 variants rs1006737, (p= $7.0\text{x}10^{-8}$) and rs1024582 (p= $1.7\text{x}10^{-7}$) of CACNA1C showed strong association with BD in their GWAS study (29). In the same year, Sklar and colleagues also detected initial signal for rs1006737 in their GWAS though they could not replicate the signal in their replication study (44).

In 2011, a large-scale GWAS using 11 sets of individual samples (covering the samples used in the two GWAS by Ferreira and Sklar in 2008) once again confirmed the genome-wide significant association of CACNA1C with BD [45]. The SNP found in this large-scale GWAS was rs4765913 (p= $1.52\text{x } 10^{-9}$).

In the GWAS done by Green and colleagues, they were able to detect an association of SNP rs1006737 with their new UK sample (22). It is worth noticing that CACNA1C is also implicated in schizophrenia. GWAS combining BD samples and schizophrenia samples produced a strong signal for association of CACNA1C (rs4765905, $p= 1.23\times10^{-8}$) [46]. Apart from GWAS, the allele "A" on SNP rs1006737 of CACNA1C was found to be associated to psychotic BD (p=0.017) in the family-based study by Lett (36). Indeed schizophrenia and BD have some similar symptoms that implied there may be certain common genetic basis contributed to both diseases [47]. In fact, a large-scale GWAS with the aim of identifying common genetic elements shared by 5 major psychiatric disorders was carried out [48]. These disorders include bipolar disorder (BD), schizophrenia (SCZ), major depressive disorder (MDD), autism (ASD) and attention deficit-hyperactivity disorder (ADHD). The study identified an association of the CACNA1C variant (rs1024582, $p=1.87x10^{-8}$) and a CACNB2 variant (rs2799573, $p=4.29\times10^{-8}$) with genome-wide significance in about 33,000 cases compared with near 28,000 controls all with European ancestry. Further investigation on the relationship between CACNA1C and the shared symptoms perhaps can provide a clearer picture to the pathophysiology of psychiatric diseases, and this approach may be potentially better than studying them as separate diseases. While it remains unclear how these genetic variants of CACNA1C are linked to BD or other psychiatric disorders, studies have been carried out to link CACNA1C variants to altered brain structure and brain functions [49], [50, 51]. It is worth noting that all the above SNPs found associated with BD (rs1006737, rs1024582, rs4765913 and rs4765905) lie in intron 3 of the gene [52]. One of the possible effects of these genetic variants might be altered expression. Bigos et al. found greater level expression of CACNA1C in the brain samples of homozygous carriers of "A" allele of SNP rs1006737 [53]. It is also possible that the variants influence alternative splicing. Alternative splicing is common in the human brain [54] and it has been found that there is a diversity of splicing variants for CACNA1C in the brain [55]. Further investigations would shed light on if the variants indeed affect this aspect.

Besides strong signal of CACNA1C, other subunits of the calcium channel are also found associated with BD. The GWAS by Psychiatric GWAS Consortium Bipolar Disorder Working Group in 2011 also detected signals with CACNB3 (rs11168751, p=2.51x10-6) and CACNA1D (rs3774609, p= 3.73x10-5) though both associations did not reach genome-wide significance (45). CACNB3 codes another subunit - beta 3 subunit of the voltage-dependent calcium channel while CACNA1D encodes the alpha 1D subunit of a voltage-dependent calcium channel, an isoform of alpha 1C subunit. Association of CACNB2 with BD was seen in Asian population, suggested by Lee and colleagues in their GWAS using Han BPI patients [35]. Two SNPs were identified: rs2073831 ($p=9.74 \times 10^{-6}$) and rs11013860 ($p=5.15 \times 10^{-5}$). The former SNP lies near KCTD12 and the latter one lies in the intronic region of CACNB2. CACNB2 codes for yet another subunit of the calcium channel whereas KCTD codes for auxiliary subunit of GABA-B receptors which can in turn regulate voltage-dependent calcium channels [56].

Genes composing another subclass of GABA receptors – the ionotropic GABA-A receptors was also implicated in BD in some studies. Association of SNP rs7680321 ($p=6.2x10^{-5}$) lying in the intronic region of GABRB1 was detected in a GWAS by The Wellcome Trust Case Control Consortium (WTCCC) [57]. Further analysis of the WTCCC data suggested a link of GABRB1 to schizoaffective disorder, bipolar type [58]. Evidence of association of GABRB2 with BD was found in a family-based association study as well [59] and this gene too has been implicated in schizophrenia [60].

CACNA1C, CACNA1D, CACNB2 and CACNB3 code for different subunits of the L-type voltage-dependent calcium channel while GABA receptors can interact with calcium channels indirectly [56, 61[. All these results point to the possibility that the L-type voltage-dependent calcium channel may be a shared genetic component in major mood disorders including BD. This may explain the high comorbidity in BD with other mood disorders [62] and a higher incidence of mood disorder in the offspring of BD patients [63].

Functions of ANK3 and CACNA1C Related to Ion Channelopathy

The formal gene name of ANK3 is Ankyrin G, node of Ranvier and it codes for protein Ankyrin G. The neural specific isoform Ankyrin G is found in locations with abundant sodium channels such as nodes of Ranvier and axon initial segments the of a neuron [64]. Ankyrin G acts as a linker protein between certain membrane proteins and the cytoskeleton. These membrane proteins include voltage-dependent sodium channel, potassium channels and cell adhesion molecule like neurofascin. The sodium and potassium channels are critical for the generation of action potential while neurofascin maintains neuronal stability [65]. Ankyrin G is important for neuronal functions probably by organizing these molecules to specific locations at a certain time [66]. Anomaly in ankyrin G may thus deregulate the membrane proteins, disrupting their normal functions. As mentioned above, reduced ankyrin G level was found in the brain tissues bipolar patient and caused mood disruption in mice [26, [41]. Though there seems to be no evidence indicating that ankyrin G also regulates the voltage-dependent calcium channel currently. CACNA1C and CACNA1D code for two isoforms of α_1 subunit of the L-type calcium channel - $Ca_v1.2$ and Cav1.3 respectively. The two isoforms are expressed in neurons but also in many other cell types. A study using Cav1.2 and Cav1.3 gene-knockout mice suggested these two isoforms regulated the mood behavior in mice [67]. Neuronal L-type calcium channels are also known to be involved in signaling cascades including MAPK pathway and cAMP signalling pathway leading to the activation of CREB [68]. CREB is a transcription factor which can also induce expression BDNF and circadian genes [69]. The signaling pathways related to BD will be discussed in the next session.

Nonetheless, from what have been discussed so far both ankyrin G and calcium channels seem promising targets for the study of BD. But these two candidates may not necessarily be inter-related for their roles in BD. Although ANK3 has been implicated in schizophrenia in some studies [38, 70], other studies contradicted the association of ANK3 with mood disorders other than BD [37, 71]. In contrast, the association of calcium channel genes in multiple mood disorders was supported by numerous studies [52]. Results in a recent GWAS (combining samples of 5 major psychiatric disorders) which found strong associations of calcium channel genes but not ANK3 [48] also support this view. It is possible that the two components contribute to different phenotypes of BD.

Candidate Genes in Signaling Pathways: BDNF, NTRK2, DGKH and NRG1

It is widely accepted that BD is a polygenic and heterogeneous disorder. Association studies often yield conflicting results and each candidate gene found so far seems to contribute only to a small risk towards BD. That has led to the ideas of epistasis and cumulative effect of multiple genes in BD [72]. And given the fact that the drugs used in treating BD, primarily lithium, can interfere with different signaling pathways [73, 74], some began to suspect that disruption in the signaling pathways in the neurons may be the underlying reason for BD symptoms. These signaling pathways include signaling pathways mediated by GSK3 (Glycogen synthase kinase-3) and G-protein coupled pathways through either PKC (Protein kinase C) or PKA (Protein kinase A). Lithium is a known inhibitor for GSK3 and IMPase [75]. Overview of the pathways in connection with BD can be found in Figure1.

BDNF and NTRK2 in Neurotrophin Signaling Pathway

GSK3 is an important kinase that participates in numerous pathways. It is a downstream target for numerous signaling molecules, with BDNF one of them (Figure 1). BDNF i.e., brain-derived neurotrophic factor is a growth factor important for neuronal function [reviewed in [76]. BDNF has been linked to mood alterations through animal studies: reduced BDNF level was found in mice with depressive-like behavior [77] whereas injection of BDNF alleviated the depression-related symptoms [78]. Regarding human studies, BDNF level was found to be increased in depressive BD patients but decreased in manic BD patients respectively after antipsychotic treatment [79]. However, this result was contested by another study in which no substantial change of BDNF and trk-B level was observed after treatment with a mood stabilizer [80]. Still a recent study of BPI patients found elevated BDNF levels in bipolar patients and the elevation was particularly prominent in patients suffering a manic episode at the time of study [81]. Concerning genetic studies, BDNF gene was reported to be associated with BD in 2 family-based studies in 2002 [82, 43], but a subsequent large-scale case-control study only found a possible association in the rapid-cycling form of BD [83]. Studies on BDNF gene mainly focuses on one SNP (rs6265) that causes valine-to-methionine substitution in the amino acid sequence (val66met). All three studies and a new association study [84] suggested that the valine allele is a possible risk allele in BD, but one study suggested the opposite – the methionine allele as the risk allele.

A candidate gene closely related to BDNF is NTRK2 which codes for neurotrophic tyrosine kinase receptor type 2 otherwise known as trk-B. It is the primary receptor for binding of BDNF. Reduced expression of NTRK2 was found in bipolar patients [86]. Association of NTRK2 (rs2769605, $p=4.5 \times 10^{-5}$) with BD was found in one GWAS involving individuals with African ancestry [87]. Studies on the association of NTRK2 as well as BDNF gene with drug response have been carried out to assess their roles in BD [88].

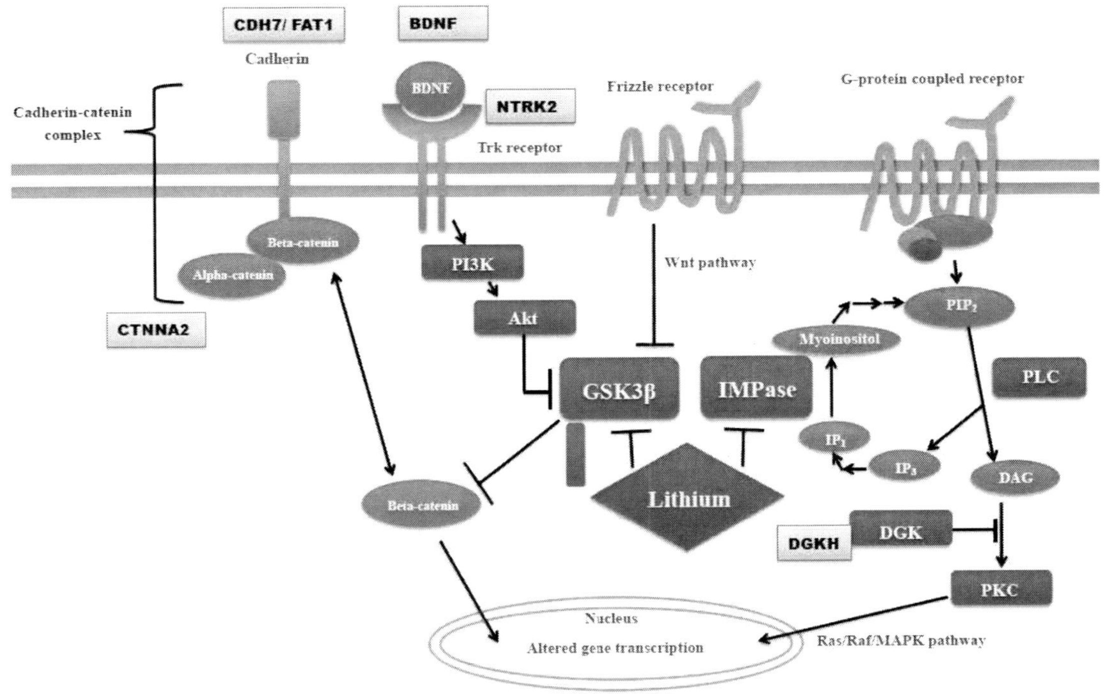

Figure 1. Overview of the signaling pathways in connection with BD with the implicated genes labeled in purple. From left to right: Cadherin including FAT1 and CDH7 proteins interact with beta-catenin. Complex formation between cadherin and caterin may hinder the Wnt signaling pathway. The neurotropin signaling pathway is an important pathway for neuronal survival and functions. Binding of BDNF to Trk receptor (encoded by NTRK2) can ultimately cause inibition of GSK3. GSK3 is a shared component in many different pathways including the neurotropin pathway and Wnt signaling pathway. it also interacts with the cadherin pathway by phosphorylating beta-catenin hence subjecting it to degradation. Lithium, the major drug for trating BD is known inhibitor of GSK3 and IMPase. IMPase plays a role in PLC/PKC signaling pathway. Inhibition of IMPase inhibition interferes with PI synthesis and possibly the PLC/PKC pathway. One gene on this pathway show possible association with BD:DGKH, DGKH encodes an isoform of DGK which phosphirylates DAG in the PLC/PKC pathway. Such phosphorylation interrupts the PLC/PKC pathway because DAG after phosphorylaion will not bind to PKC.

DGKH in G-Protein Coupled Signaling Pathways

DGKH encodes one of the isoforms of diacylglycerol kinase (DGK). The enzyme phosphorylates diacylglycerol (DAG) which activates PKC for the continuation of the signal cascade. Therefore, the enzyme can inhibit the signaling transduction through PKC pathway. DGKH gene was first implicated in the etiology of BD in a GWAS done by Baum and colleagues in 2008 [90]. In this GWAS using BPI sample of European ancestry, strong association of one SNP rs1012053 ($p=1.5x\ 10^{-8}$) in the intronic region of DGKH was detected. Another GWAS studying Bulgarian population detected association of the same SNP with only modest significance ($p=4.50 \times 10^{-2}$) in terms of a GWAS study [91]. Two association studies with Finnish BPI sample and Sardinian BPI sample respectively identified the association of SNP rs9315885 (both $p< 0.05$) [92, 93]. Zeng and colleagues found an association of another SNP rs2122246 ($p=0.0186$) with BPI subgroup in their Chinese Han samples [94].

Weber and colleagues studied the DGKH gene association with BD and two other psychiatric disorders: namely MDD and adult ADHD [95]. They found association of two SNPs rs1170169 ($p=0.013$) and rs9525580 ($p=0.007$) with BD. Association of the rs9525580 was also detected with MDD and ADHD (both $p<0.05$). However, one association study with the Scandinavian sample did not found strong association of DGKH with BD [96]. It is worth noticing that BPII patients were also included in their study, in contrast to the 3 other association studies by Ollila, Squassina and Zeng [92, 93, 94]. In another association study with Japanese samples with both BPI and BPII, there was no association found but from a meta-analysis adding the sample from Zeng's study [94], one SNP rs9315897 ($p= 0.039$) was found associated with BD [97]. Hence it is possible that DGKH gene is a risk factor specific to certain subtype of BD (BPI) or there are different risk alleles of DGKH that are specific to certain populations. There is one expression study of DGKH in BD to date. An increase in mRNA level of DGKH was found in the post-mortem brain tissues from bipolar patients [98]. More studies are needed to confirm the involvement of DGKH in BD and how it exerts its effects.

NRG1

Neuregulin 1 is a signaling protein that acts through tyrosine kinase receptors. It is also highly expressed in brain tissues and is important for the development of the nervous system [reviewed in [99]. Several association studies suggested the association of NRG1 with BD [100, 101, 102] though the association appeared to be limited to psychotic BD. Two other association studies, which did not separate BD samples into psychotic subsets, could not find any association between NRG1 and BD [103, 104].

Different genetic variants in signaling pathways, each contributing to a small effect, combine to make a significant impact to cause the disease, This model explains the relatively low odds ratios for individual genetic variants and the difficulty in replicating the results of previous studies. Considering this, several studies have been carried out to search for the connection between different signaling pathways and BD [72, 105, 106]. However, given the complexity of the signaling pathways and the great extent they cover, there is a need to pinpoint those components responsible for the etiology of BD, if they exist. Signal transduction is a vital process for virtually all cells and most of the signaling pathways, with the exception of neurotrophin pathway, are not exclusive to brain cells. Disruption of the signaling pathways only in neurons would suggest that problems occurred in brain-specific isoforms of the members in the signaling pathways. Thus it would be necessary to identify the relevant isoforms. Another point worth noticing is that studying the working mechanisms of the drug for the treatment of BD, lithium, for example, may help us to locate those components responsible for the symptoms of BD. But given that each signaling pathway trigger various responses downstream thus influencing many aspects of the cellular activities, identifying the ones that control mood change may be difficult. For instance, although lithium is known to inhibit GSK-3, GSK-3 is involved in various important intracellular signaling pathways controlling cellular proliferation and even cell survival. Incorporating the gene candidates discovered in GWAS – studying their roles in the signaling pathways might help narrow the search.

Cell Adhesion Molecules and BD: FAT1, CTNNA2, CDH7 and NCAN

Cell adhesion molecules are important for cell-cell interactions and some of them are also involved in signaling pathways, both of which are crucial for neuronal functions. Several genes involving cell adhesion have been implicated in BD. They are FAT1, CTNNA2, CDH7 and NCAN.

FAT1 codes for an integral membrane protein which belongs to the cadherin superfamily. FAT1 was proposed to be related to BD by the study conducted by Blair and colleagues [107]. Their result was later replicated by a case-control study in which a new non-synonymous SNP rs2637777 ($p= 2.8x\ 10^{-4}$) was detected [108]. The association of the FAT1 gene was also supported by GWAS but the significance did not reach genome-wide significance (rs455219, $p=1.5x10^{-4}$) [109].

In the study done by Blair and colleagues (107), they investigated the association of FAT1 in BD with samples from Australia, UK and Bulgarian. Most Australian subjects are of British or Irish origin. Hence a pooled analysis with Australian and UK samples were performed. Strong associations were found in two SNPs rs2304865 ($p= 2x\ 10^{-4}$) and rs1298865 ($p=3x10^{-3}$). The former SNP lies in the intronic region of FAT1 while the latter one causes synonymous substitution in mRNA of FAT1. The SNP used in the GWAS rs455219 also lies in the intronic region of FAT1. Thus these SNPs appear to be markers for FAT1, but not the functional variant of FAT1 that directly contribute to BD. Indeed Blair and colleagues found no difference in FAT 1 mRNA level in the post-mortem brain samples from BD patients and healthy controls in the expression analysis. But in the follow-up animal study, they found that chronic administration of lithium (7 days) caused a decrease in FAT1 mRNA level but an increase in CATNB mRNA level in mice. CATNB gene codes for cadherin association protein – beta-catenin, a protein that is important in cell adhesion but also in Wnt/GSK3 signaling pathway (Figure 1). A recent study reported that FAT1 protein can interact with beta-catenin leading to the inhibition of Wnt signaling pathway [110]. Together the two studies seem to provide more evidence in the possible involvement of the signaling pathway in BD. But it should be pointed out that Wnt signaling pathway is crucial in many developmental events. Gene mutations in this pathway are known to cause developmental abnormalities or contribute to cancers [reviewed in [111]]. In comparison, the symptoms in BD fluctuate between two extreme states and in between patients can appear to be normal (asymptomatic) for a long period of time i.e., in remission. One would need to address how genetic variations in signaling pathways could cause such fluctuations rather than permanent changes in cellular activities. In addition, the functions of the Wnt pathway in the adult brain cells remain unclear and thus more studies are needed to confirm the role of Wnt pathway in BD.

Another member in the catenin family was also reported to be associated with BD. Scott and colleagues identified one SNP rs13409348 ($p=2.7x10^{-6}$) in the intronic region of CTNNA2 in GWAS of BD [34]. CTNNA2 were also reported to be associated with excitement seeking (rs7600563, $p= 2.04x\ 10^{-8}$) [112], schizophrenia [113] and adult ADHD [114]. CTNNA2 is expressed in neurons and it codes for alpha-2-catenin. The protein interacts with N-cadherin to form a complex important in regulating synaptic plasticity. Alpha-catenin can also bind beta-catenin, making it possible to interfere with Wnt signaling pathway.

CDH7 is another cadherin gene to be implicated in BD. Association of CDH7 with BD was first identified in a GWAS in 2008 (44). In that study, 6 SNPs representing CDH7 gene were found to be associated with BD ($p= 1.83x10^{-5} – 3.26x10^{-5}$) (refer to Table 1) but none of these SNPs lies on the gene region. Soronen and colleagues successfully replicated the association of CDH7 (p=0.012- 0.044) (refer to Table 1) with BD using a Finnish sample.

NCAN which encodes for neurocan was also suspected to play a role in the pathogenesis of BD. Neurocan is a proteoglycan in the extracellular matrix that modulates cell adhesion. A recent GWAS found a strong association of SNP rs1064395 ($p= 3.02x10^{-8}$) with BD [116]. This SNP is located in the 3' untranslated region of NCAN. Higher frequency of the minor allele "A" was observed in BD patients in this study. A subsequent study showed NCAN knockout mice exhibited manic-like behavior [117] suggested NCAN is a promising candidate for future studies of BD.

Animal Studies Implicated Circadian Genes in BD

It has been hypothesized the circadian rhythm genes are involved in BD. This is because sleep disturbance is one of the major symptoms found in BD patients. BD patients during a manic episode often experience difficulty in sleeping at night while during a depressive episode the situation reverses, i.e., they tend to over-sleep. Several circadian rhythm genes like CLOCK, TIMELESS and ARNNTL1 have been implicated in BD in some association studies [118, 119] but there are also studies that disagreed with the association [120, 121]. The majority of the studies on circadian genes are animal studies. For example, several studies showed CLOCK knockout mice exhibited mania-like behavior (122) (123) (124). However little evidence has been found in the GWAS of BD, none of the circadian genes has been detected so far. As mentioned before, calcium channel signaling interacts with circadian cycles [69]. It may be possible that genetic variance in those upstream components in the pathway ultimately causes the disruption of circadian rhythms.

FINDINGS IN EPIGENETIC STUDIES

Epigenetics refers to the study of heritable changes without changes of DNA sequences. Examples of epigenetic processes include DNA methylation, histone modification, chromatin remodeling and so on. One important difference between epigenetics and genetics is that certain epigenetic status such as DNA methylation can change at different ages, different tissues or in response to external stimuli [125], so the epigenome is more dynamic. Epigenetics may be one of the reasons for the discordance in monozygotic bipolar twins. Recent epigenetic studies of BD mainly focus on searching epigenetically modified genes in BD individuals for better understanding of the etiology of BD. Some studies also tried to investigate the machinery of epigenetic modification including the enzymes and cofactors necessary for epigenetic modifications.

Genes with Altered Methylation Patterns in BD Patients

Differential methylation patterns have been found in numerous genes. Most of these results came from twin studies while some being case-control studies. As mentioned above, the epigenome is more dynamic than the genome so it is more difficult to locate the real target. Compared with genetic study, the epigenetic study of BD is rather a new area of research. Therefore the number of findings for epigenetic studies is much smaller than that of genetic studies.

Other Genes Implicated Also in Genetic Studies HTR1A and SLC6A4

HTR1A encodes G protein-coupled serotonin receptor 1A. HTR1A has been implicated in BD by some association studies but the results were not conclusive [126, 127]. Concerning the epigenetic aspect, Carrard and colleagues found increased methylation in 5HTRA in the BD patients [128]. However blood sample but not brain tissue was used in this study. Since DNA methylation patterns vary among different tissues, future studies with brain sample would be needed to confirm the link BD to this methylation difference. SLC6A4 codes for a serotonin transporter in the neurons. A study by Sugawara and colleagues found hypermethylation in SLC6A4 in BD patients when compared with their discordant twins. This result was supported by DNA methylation analysis of postmortem brain tissues from another set of BD patients [129]. A subsequent study by the same team found that known mood stabilizers i.e., lithium, valproate and carbamazepine decrease the degree of methylation in 2 CpG sites of SLC6A4.

COMT

COMT codes for catechol-O-methyltransferase, enzyme responsible for degrading dopamine. Several association studies investigated the relationship between COMT variants and phenotypic differences in patients [131, 132]. Study of post-mortem brain tissues showed hypomethylation in the promoter region of COMT in BD patients when compared with controls. Further analysis also found the corresponding increase in mRNA level of MB-

COMT (membrane bound form of catechol-O-methyltransferase which is the major form found in brain) in BD patients [133]. In 2011, the same research team successfully replicated the result using saliva samples [134]. This suggested epigenetic modification of COMT was not tissue-specific (or they appear to be similar in saliva and brain).

CTNNA2 and MCHR1

Both CTNNA2 and MCHR1 were found hypomethylated in bipolar subjects in a discordant twin study [135]. MCHR1 was also implicated in BD and schizophrenia by one gene association study [136]. MCHR1, also known as GPR24, encodes melanin-concentrating hormone receptor 1. Studies showed antagonists of the receptor possessed antidepressant effect which implied MCHR1 might play a role in mood regulation [137, 138]. Another interesting finding in this study was that for ZNF659 gene, hypermethylation was observed in schizophrenia subjects while hypomethylation was observed in BD subjects, when compared with that of their healthy twin brother/sister.

Non-Protein Coding Gene: HCG9

The official name of HCG9 is HLA complex group 9 which is located in MHC class I region. HCG9 does not encode a protein and function of its gene product is still unknown. Hypomethylation of HCG9 in BD patients was found in a multi-tissue analysis [139]. Hypomethylation of the gene was found in post-mortem brain tissues, white blood cells as well as sperm samples from different sources. The consistency in the methylation difference in various tissues led researchers to suggest the use of easily accessed cells (e.g., white blood cells) as sources for finding a potential biomarker for BD.

Investigation on the Machinery of Epigenetic Modification

Numerous studies showed altered DNA methyltransferase (DNMT) expression in BD patients. There was an increase in DNMT1 mRNA level in the postmortem brain tissues of BD patient when compared with that of the controls [140]. Besides methyltransferase, the same research team also studied S-adenosyl methionine levels in patients with mental illness [141]. S-adenosyl methionine is a cofactor which serves as a methyl donor in methylation reactions. S-adenosyl methionine converts to S-adenosyl homocysteine after donating the methyl group. The study by Guidotti and colleagues found an increase in both S-adenosyl methionine and S-adenosyl homocysteine levels in the post-mortem brain tissues of BD patients. Such result might suggest a greater degree of DNA methylation but it should also be pointed out that S-adenosyl methionine is a universal methyl donor meaning it also donates its methyl group to many other molecules. In a recent study on DNMT1 expression, decreased mRNA level of DNMT1 was found in white blood cells of BD patient suffering a depressive episode. In contrast, such decrease was not observed in BD patients in remission. [142]. This suggests fluctuation of DNMT1 may be related to the mood shifts in BD patients.

CONCLUSION

Bipolar disorder is a complex psychiatric disorder. Even though with high heritability found in concordance studies, it is now clear that it is a polygenic disease. There are a vast number of candidate genes in bipolar disorder and only a few can be covered in this review. Nevertheless from the findings mentioned above, we can see multiple interactions among these genes which make it more difficult to identify the real targets. Perhaps collaboration with studies of other diseases would help understanding the etiology of bipolar disorder. It is already clear that there are genetic overlaps between bipolar disorder and other psychiatric diseases such as schizophrenia and depression. The common risk genes "shared" by various diseases allow us to link them to specific phenotypes that are also shared by the diseases. On the other hand, it would be logical to deduce that genes found exclusive to one particular disease that are likely to contribute to phenotypes specific to that disease. As for the study of bipolar disorder, one distinctive feature of this disorder that set it apart from other psychiatric disorders is the mood shift. Identifying the "switch" of this mood change would offer greater insight into the etiology of bipolar disorder. Therefore, studying the links between genes and specific phenotypes or endophenotypes may prove more useful than studying the link with the overall status of the patient. In addition, more adoption studies using the current diagnostic criteria may be carried out to help distinguish the genetic components from the environmental components in the etiology of bipolar disorder.

Summary Points

- Bipolar disorder is a devastating mental illness with a large genetic component.
- Different study approaches have advantages and limitations with none being the best approach.
- Genome-wide association study in bipolar disorder lead to discovery of new candidate genes and support for channelopathy.
- Different signaling pathways may be disrupted in bipolar disorder.
- Circadian rhythm is disrupted in bipolar patients but circadian genes may not necessarily be the cause.
- Epigenetic study of bipolar disorder is a rather new topic but initial results show a convergence with the genetic studies.

Future Issues

- Cross-disease study in hope of more insights on the etiology of bipolar disorder.
- Association of genes to specific phenotypes.
- Identification of genes that affect treatment response of bipolar disorder.
- Use of iPS stem cells to study bipolar disorder.
- Use of high-throughput sequencing to find rare variants with high impact of genetic risk.

REFERENCES

[1] Merikangas KR, Jin R, He JP, Kessler RC, Lee S, et al. Prevalence and Correlates of Bipolar Spectrum Disorder in the World Mental Health Survey Initiative. *Arch Gen Psychiatry.* 2011 March; 68(3): p. 241-51.

[2] Kleinman NL, Brook RA, Rajagopalan K, Gardner HH, Brizee TJ, et al. Lost time, absence costs, and reduced productivity output for employees with bipolar disorder. *J. Occup. Environ. Med.* 2005 November; 47(11): p. 1117-24.

[3] Bruffaerts R, Vilagut G, Demyttenaere K, Alonso J, Alhamzawi A, et al. Role of common mental and physical disorders in partial disability around the world. *Br. J. Psychiatry.* 2012 June; 200(6): p. 454-61.

[4] Parker G, McCraw S, Hadzi-Pavlovic D, Fletcher K. Costs of the principal mood disorders: A study of comparative direct and indirect costs incurred by those with bipolar I, bipolar II and unipolar disorders. *J. Affect. Disord.* 2013 July; 149(1-3): p. 46-55.

[5] Fajutrao L, Locklear J, Priaulx J, Heyes A. A systematic review of the evidence of the burden of bipolar disorder in Europe. *Clin Pract Epidemiol Ment Health.* 2009 January; 5: p. 3.

[6] Young AH, Rigney U, Shaw S, Emmas C, Thompson JM. Annual cost of managing bipolar disorder to the UK healthcare system. *J. Affect. Disord.* 2011 October; 133(3): p. 450-6.

[7] Begley CE, Annegers JF, Swann AC, Lewis C, Coan S, et al. The lifetime cost of bipolar disorder in the US: an estimate for new cases in 1998. *Pharmacoeconomics.* 2001; 19: p. 483-95.

[8] Dilsaver SC. An estimate of the minimum economic burden of bipolar I and II disorders in the United States: 2009. *J. Affect. Disord.* 2011 march; 129(1-3): p. 79-83.

[9] Lichtenstein P, Yip BH, Bjork C, Pawitan Y, Cannon TD, et al. Common genetic determinants of schizophrenia and bipolar disorder in Swedish families: a population-based study. *Lancet* . 2009 January; 373(9659): p. 234-9.

[10] Wray NR, Gottesman II. Using summary data from the Danish national registers to estimate heritabilities for schizophrenia, bipolar disorder, and major depressive disorder. *Front. Genet.* 2012 July; 3: p. 118.

[11] McGuffin P, Rijsdijk F, Andrew M, Sham P, Katz R, et al. The heritability of bipolar affective disorder and the genetic relationship to unipolar depression. *Arch Gen Psychiatry.* 2003 May; 60(5): p. 497-502.

[12] Kieseppä T, Partonen T, Haukka J, Kaprio J, Lönnqvist J. High concordance of bipolar I disorder in a nationwide sample of twins. *Am J Psychiatry.* 2004 October; 161(10): p. 1814-21.

[13] Edvardsen J, Torgersen S, Røysamb E, Lygren S, Skre I, et al. Heritability of bipolar spectrum disorders. Unity or heterogeneity? *J. Affect. Disord.* 2008 Mar; 106(3): p. 220-40.

[14] Gottesman II, Laursen TM, Bertelsen A, Mortensen PB. Severe mental disorders in offspring with 2 psychiatrically ill parents. *Arch Gen Psychiatry.* 2010 Mar; 67(3): p. 252-7.

[15] Frazer KA, Murray SS, Schork NJ, Topol EJ. Human genetic variation and its contribution to complex traits. *Nat Rev Genet.* 2009 Apr; 10(4): p. 241-51.
[16] McQueen MB, Devlin B, Faraone SV, Nimgaonkar VL, Sklar P, et al. Combined analysis from eleven linkage studies of bipolar disorder provides strong evidence of susceptibility loci on chromosomes 6q and 8q. *Am J Hum Genet.* 2005 Oct; 77(4): p. 582-95.
[17] Zandi PP, Zoellner S, Avramopoulos D, Willour VL, Chen Y, et al. Family-Based SNP Association Study on 8q24 in Bipolar Disorder. *Am J Med Genet B Neuropsychiatr Genet.* 2008 Julyl; 147B(5): p. 612-8.
[18] Fan J, Ionita-Laza I, McQueen MB, Devlin B, Purcell S, et al. Linkage disequilibrium mapping of the chromosome 6q21–22.31 bipolar I disorder susceptibility locus. *Am J Med Genet B Neuropsychiatr Genet.* 2010 Jan; 153B(1): p. 29-37.
[19] Kato T, Iwamoto K, Kakiuchi C, Kuratomi G, Okazaki Y. Genetic or epigenetic difference causing discordance between monozygotic twins as a clue to molecular basis of mental disorders. *Mol Psychiatry.* 2005 July; 10(7): p. 622-30.
[20] Bell JT, Spector TD. A twin approach to unraveling epigenetics. *Trends Genet.* 2011 Mar; 27(3): p. 116-25.
[21] J O. Genetic approaches to a better understanding of bipolar disorder. *Pharmacol Ther.* 2012 Feb; 133(2): p. 133-41.
[22] Green EK, Hamshere M, Forty L, Gordon-Smith K, Fraser C, et al. Replication of bipolar disorder susceptibility alleles and identification of two novel genome-wide significant associations in a new bipolar disorder case-control sample. *Mol Psychiatry.* 2013 Dec; 18(12): p. 1302-7.
[23] Rao JS, Kellom M, Reese EA, Rapoport SI, Kim HW. Dysregulated glutamate and dopamine transporters in postmortem frontal cortex from bipolar and schizophrenic patients. *J Affect Disord.* 2012 Jan; 136(1-2): p. 63-71.
[24] Myles-Worsley M, Tiobech J, Browning SR, Korn J, Goodman S, et al. Deletion at the SLC1A1 glutamate transporter gene co-segregates with schizophrenia and bipolar schizoaffective disorder in a 5-generation family. *Am J Med Genet B Neuropsychiatr Genet.* 2013 Mar; 162B(2): p. 87-95.
[25] Kakefuda K, Oyagi A, Ishisaka M, Tsuruma K, Shimazawa M, et al. Diacylglycerol kinase β knockout mice exhibit lithium-sensitive behavioral abnormalities. *PLoS One.* 2010 Oct; 18(5).
[26] Leussis MP, Berry-Scott EM, Saito M, Jhuang H, de Haan G, et al. The ANK3 bipolar disorder gene regulates psychiatric-related behaviors that are modulated by lithium and stress. *Biol Psychiatry.* 2013 Apr; 73(7): p. 683-90.
[27] Takata A, Kim SH, Ozaki N, Iwata N, Kunugi H, et al. Association of ANK3 with bipolar disorder confirmed in East Asia. *Am J Med Genet B Neuropsychiatr Genet.* 2011 Apr; 156B(3): p. 312-5.
[28] Schulze TG, Detera-Wadleigh SD, Akula N, Gupta A, Kassem L, et al. Two variants in Ankyrin G (ANK3) are independent genetic risk factors for bipolar disorder. *Mol Psychiatry.* 2009 May; 14(5): p. 487-91.
[29] Ferreira MA, O'Donovan MC, Meng YA, Jones IR, Ruderfer DM, et al. Collaborative genome-wide association analysis supports a role for ANK3 and CACNA1C in bipolar disorder. *Nat Genet.* 2008 Sep; 40(9): p. 1056-8.

[30] Seifuddin F, Mahon PB, Judy J, Pirooznia M, Jancic D, et al. Meta-analysis of genetic association studies on bipolar disorder. *Am J Med Genet B Neuropsychiatr Genet*. 2012 Jul; 159B(5): p. 508-18.

[31] Lee KW, Woon PS, Teo YY, Sim K. Genome wide association studies (GWAS) and copy number variation (CNV) studies of the major psychoses: what have we learnt? *Neurosci Biobehav Rev*. 2012 Jan; 36(1): p. 556-71.

[32] R F. Channelopathies: ion channel defects linked to heritable clinical disorders. *J Med Genet*. 2000 Oct; 37(10): p. 729-40.

[33] Gargus JJ. Ion channel functional candidate genes in multigenic neuropsychiatric disease. *Biol Psychiatry*. 2006 Jul; 60(2): p. 177-85.

[34] Scott LJ, Muglia P, Kong XQ, Guan W, Flickinger M,et al. Genome-wide association and meta-analysis of bipolar disorder in individuals of European ancestry. *Proc Natl Acad Sci U S A*. 2009 May; 106(18): p. 7501-6.

[35] Lee MT, Chen CH, Lee CS, Chen CC, Chong MY, et al. Genome-wide association study of bipolar I disorder in the Han Chinese population. Mol Psychiatry. 2011 May; 16(5): p. 548-56.

[36] Lett TA, Zai CC, Tiwari AK, Shaikh SA, Likhodi O, et al. ANK3, CACNA1C and ZNF804A gene variants in bipolar disorders and psychosis subphenotype. *World J Biol Psychiatry*. 2011 Aug; 12(5): p. 392-7.

[37] Gella A, Segura M, Durany N, Pfuhlmann B, Stöber G, et al.. Is Ankyrin a genetic risk factor for psychiatric phenotypes?. *BMC Psychiatry*. 2011 Jun; 11(103).

[38] Athanasiu L, Mattingsdal M, Kähler AK, Brown A, Gustafsson O, et al. Gene variants associated with schizophrenia in a Norwegian genome-wide study are replicated in a large European cohort. *J Psychiatr Res*. 2010 Sep; 44(12): p. 748-53.

[39] Ruberto G, Vassos E, Lewis CM, Tatarelli R, Girardi P, and et al. The cognitive impact of the ANK3 risk variant for bipolar disorder: initial evidence of selectivity to signal detection during sustained attention. *PLoS One*. 2011 Jan; 6(1).

[40] Roussos P, Giakoumaki SG, Georgakopoulos A, Robakis NK, Bitsios P. The CACNA1C and ANK3 risk alleles impact on affective personality traits and startle reactivity but not on cognition or gating in healthy males. *Bipolar Disord*. 2011 May; 13(3): p. 250-9.

[41] Rueckert EH, Barker D, Ruderfer D, Bergen SE, O'Dushlaine C, et al. Cis-acting regulation of brain-specific ANK3 gene expression by a genetic variant associated with bipolar disorder. *Mol Psychiatry*. 2013 Aug; 18(8): p. 922-9.

[42] Dolphin AC. Calcium channel diversity: multiple roles of calcium channel subunits. *Curr Opin Neurobiol*. 2009 Jun; 19(3): p. 237-44.

[43] Sklar P, Gabriel SB, McInnis MG, Bennett P, Lim YM, et al. Family-based association study of 76 candidate genes in bipolar disorder: BDNF is a potential risk locus. Brain-derived neutrophic factor. *Mol Psychiatry*. 2002; 7(6): p. 579-93.

[44] Sklar P, Smoller JW, Fan J, Ferreira MA, Perlis RH, et al. Whole-genome association study of bipolar disorder. *Mol Psychiatry*. 2008 Jun; 13(6): p. 558-69.

[45] Psychiatric GWAS Consortium Bipolar Disorder Working Group. Large-scale genome-wide association analysis of bipolar disorder identifies a new susceptibility locus near ODZ4. *Nat Genet*. 2011 Sep; 43(10): p. 977-83.

[46] Hamshere ML, Walters JT, Smith R, Richards AL, Green E, et al. Genome-wide significant associations in schizophrenia to ITIH3/4, CACNA1C and SDCCAG8, and

extensive replication of associations reported by the Schizophrenia PGC. *Mol Psychiatry*. 2013 Jun; 18(6): p. 708-12.
[47] Craddock N, Owen MJ. The beginning of the end for the Kraepelinian dichotomy. *Br J Psychiatry*. 2005 May; 186: p. 364-6.
[48] Cross-Disorder Group of the Psychiatric Genomics Consortium; Genetic Risk Outcome of Psychosis (GROUP) Consortium. Identification of risk loci with shared effects on five major psychiatric disorders: a genome-wide analysis. *Lancet*. 2013 Apr; 381(9875): p. 1371-9.
[49] Perrier E, Pompei F, Ruberto G, Vassos E, Collier D, et al. Initial evidence for the role of CACNA1C on subcortical brain morphology in patients with bipolar disorder. *Eur Psychiatry*. 2011 Apr; 26(3): p. 135-7.
[50] Dima D, Jogia J, Collier D, Vassos E, Burdick KE, et al. Independent modulation of engagement and connectivity of the facial network during affect processing by CACNA1C and ANK3 risk genes for bipolar disorder. *JAMA Psychiatry*. 2013 Dec; 70(12): p. 1303-11.
[51] Erk S, Meyer-Lindenberg A, Linden D, Lancaster T, Mohnke S, et al. Replication of brain function effects of a genome-wide supported psychiatric risk variant in the CACNA1C gene and new multi-locus effects. *Neuroimage*. 2014 Mar; In Press.
[52] Bhat S, Dao DT, Terrillion CE, Arad M, Smith RJ, et al. CACNA1C (Ca(v)1.2) in the pathophysiology of psychiatric disease. *Prog Neurobiol*. 2012 Oct; 99(1): p. 1-14.
[53] Bigos KL, Mattay VS, Callicott JH, Straub RE, Vakkalanka R, et al. Genetic variation in CACNA1C affects brain circuitries related to mental illness. *Arch Gen Psychiatry*. 2010 Sep; 67(9): p. 939-45.
[54] Yeo G, Holste D, Kreiman G, Burge CB. Variation in alternative splicing across human tissues. *Genome Biol*. 2004; 5(10): p. R74.
[55] Tang ZZ, Liang MC, Lu S, Yu D, Yu CY, et al. Transcript scanning reveals novel and extensive splice variations in human l-type voltage-gated calcium channel, Cav1.2 alpha1 subunit. *J Biol Chem*. 2004 Oct; 279(43): p. 44335-43.
[56] Padgett CL, Slesinger PA. GABAB receptor coupling to G-proteins and ion channels. *Adv Pharmacol*. 2010; 58: p. 123-47.
[57] Wellcome Trust Case Control Consortium. Genome-wide association Study of 14,000 cases of seven common diseases and 3,000 shared controls. *Nature*. 2007 Jun; 447(7145): p. 661-78.
[58] Craddock N, Jones L, Jones IR, Kirov G, Green EK, et al. Strong genetic evidence for a selective influence of GABAA receptors on a component of the bipolar disorder phenotype. Mol Psychiatry. 2010 Feb; 15(2): p. 146-53. *Erratum in: Mol Psychiatry*. 2010 Nov;15(11):1121.
[59] Perlis RH, Purcell S, Fagerness J, Kirby A, Petryshen TL, et al. Family-Based Association Study of Lithium-Related and Other Candidate Genes in Bipolar Disorder. *Arch Gen Psychiatry*. 2008 Jan; 65(1): p. 53-61.
[60] Zhao C, Xu Z, Chen J, Yu Z, Tong KL, et al. Two isoforms of GABA(A) receptor beta2 subunit with different electrophysiological properties: Differential expression and genotypical correlations in schizophrenia. *Mol Psychiatry*. 2006 Dec; 11(12): p. 1092-105.

[61] Long P, Mercer A, Begum R, Stephens GJ, Sihra TS, et al. Nerve Terminal GABAA Receptors Activate Ca2+/Calmodulin-dependent Signaling to Inhibit Voltage-gated Ca2+ Influx and Glutamate Release. *J Biol Chem.* 2009 Mar; 284(13): p. 8726-37.

[62] Fagiolini A, Forgione R, Maccari M, Cuomo A, Morana B, et al. Prevalence, chronicity, burden and borders of bipolar disorder. *J Affect Disord.* 2013 Jun; 148(2-3): p. 161-9.

[63] Mesman E, Nolen WA, Reichart CG, Wals M, Hillegers MH. The Dutch Bipolar Offspring Study: 12-Year Follow-Up. *Am J Psychiatry.* 2013 May; 170(5): p. 542-9.

[64] Kordeli E, Lambert S, Bennett V. AnkyrinG. A new ankyrin gene with neural-specific isoforms localized at the axonal initial segment and node of Ranvier. *J Biol Chem.* 1995 Feb; 270(5): p. 2352-9.

[65] Kriebel M, Wuchter J, Trinks S, Volkmer H. Neurofascin: a switch between neuronal plasticity and stability. *Int J Biochem Cell Biol.* 2012 May; 44(5): p. 694-7.

[66] Leussis MP, Madison JM, Petryshen TL. Ankyrin G: genetic association with bipolar disorder and relevance to disease pathophysiology. *Biol Mood Anxiety Disord.* 2012 Oct; 2(1): p. 18.

[67] Sinnegger-Brauns MJ, Hetzenauer A, Huber IG, Renström E, Wietzorrek G, et al. Isoform-specific regulation of mood behavior and pancreatic beta cell and cardiovascular function by L-type Ca 2+ channels. *J Clin Invest.* 2004 May; 113(10): p. 1430-9.

[68] Striessnig J, Koschak A, Sinnegger-Brauns MJ, Hetzenauer A, Nguyen NK, et al. Role of voltage-gated L-type Ca2+ channel isoforms for brain function. *Biochem Soc Trans.* 2006 Nov; 34(Pt5): p. 903-9.

[69] Deisseroth K, Mermelstein PG, Xia H, Tsien RW. Signaling from synapse to nucleus: the logic behind the mechanisms. *Curr Opin Neurobiol.* 2003 Jun; 13(3): p. 354-65.

[70] Yuan A, Yi Z, Wang Q, Sun J, Li Z, et al. ANK3 as a risk gene for schizophrenia: new data in Han Chinese and meta-analysis. *Am J Med Genet B Neuropsychiatr Genet.* 2012 Dec; 159B(8): p. 997-1005.

[71] Kloiber S, Czamara D, Karbalai N, Müller-Myhsok B, Hennings J, et al. ANK3 and CACNA1C--missing genetic link for bipolar disorder and major depressive disorder in two German case-control samples. *J Psychiatr Res.* 2012 Aug; 46(8): p. 973-9.

[72] Pandey A, Davis NA, White BC, Pajewski NM, Savitz J, et al. Epistasis network centrality analysis yields pathway replication across two GWAS cohorts for bipolar disorder. *Transl Psychiatry.* 2012 Aug; 2: p. e154.

[73] Gurvich N, Klein PS. Lithium and valproic acid: parallels and contrasts in diverse signaling contexts. *Pharmacol Ther.* 2002 Oct; 96(1): p. 45-66.

[74] King JS, Teo R, Ryves J, Reddy JV, Peters O, et al. The mood stabiliser lithium suppresses PIP3 signalling in Dictyostelium and human cells. *Dis Model Mech.* 2009 May-Jun; 2(5-6): p. 306-12.

[75] Gould TD, Quiroz JA, Singh J, Zarate CA, Manji HK. Emerging experimental therapeutics for bipolar disorder: insights from the molecular and cellular actions of current mood stabilizers. *Mol Psychiatry.* 2004 Aug; 9(8): p. 734-55.

[76] Huang EJ, Reichardt LF. Neurotrophins: roles in neuronal development and function. *Annu Rev Neurosci.* 2001; 24: p. 677-736.

[77] Smith MA, Makino S, Kvetnansky R, Post RM. Stress and glucocorticoids affect the expression of brain-derived neurotrophic factor and neurotrophin-3 mRNAs in the hippocampus. *J Neurosci.* 1995 Mar; 15(3 Pt 1): p. 1768-77.

[78] Shirayama Y, Chen AC, Nakagawa S, Russell DS, Duman RS. Brain-derived neurotrophic factor produces antidepressant effects in behavioral models of depression. *J Neurosci.* 2002 Apr; 22(8): p. 3251-61.

[79] Grande I, Kapczinski F, Stertz L, Colpo GD, Kunz M, et al. Peripheral brain-derived neurotrophic factor changes along treatment with extended release quetiapine during acute mood episodes: an open-label trial in drug-free patients with bipolar disorder. *Psychiatr Res.* 2012 Nov; 46(11): p. 1511-4.

[80] Huang TL, Hung YY, Lee CT, Chen RF. Serum protein levels of brain-derived neurotrophic factor and tropomyosin-related kinase B in bipolar disorder: effects of mood stabilizers. *Neuropsychobiology.* 2012; 65(2): p. 65-9.

[81] Barbosa IG, Rocha NP, Miranda AS, Huguet RB, Bauer ME, et al. Increased BDNF levels in long-term bipolar disorder patients. *Rev Bras Psiquiatr.* 2013 Mar; 35(1): p. 67-9.

[82] Neves-Pereira M, Mundo E, Muglia P, King N, Macciardi F, et al. The brain-derived neurotrophic factor gene confers susceptibility to bipolar disorder: evidence from a family-based association study. *Am J Hum Genet.* 2002 Sep; 71(3): p. 651-5.

[83] Green EK, Raybould R, Macgregor S, Hyde S, Young AH, et al. Genetic variation of brain-derived neurotrophic factor (BDNF) in bipolar disorder: case-control study of over 3000 individuals from the UK. *Br J Psychiatry.* 2006 Jan; 188: p. 21-5.

[84] Lee SY, Chen SL, Wang YS, Chang YH, Huang SY, et al. COMT and BDNF interacted in bipolar II disorder not comorbid with anxiety disorder. *Behav Brain Res.* 2013 Jan; 237: p. 243-8.

[85] Wang Z, Li Z, Chen J, Huang J, Yuan C,et al. Association of BDNF gene polymorphism with bipolar disorders in Han Chinese population. *Genes Brain Behav.* 2012 Jul; 11(5): p. 524-8.

[86] Soontornniyomkij B, Everall IP, Chana G, Tsuang MT, Achim CL, et al. Tyrosine kinase B protein expression is reduced in the cerebellum of patients with bipolar disorder. *J Affect Disord.* 2011 Oct; 133(3): p. 646-54.

[87] Smith EN, Bloss CS, Badner JA, Barrett T, Belmonte PL, et al. Genome-wide association study of bipolar disorder in European American and African American individuals. *Mol Psychiatry.* 2009 Aug; 14(8): p. 755-63.

[88] Wang Z, Fan J, Gao K, Li Z, Yi Z, et al. Neurotrophic Tyrosine Kinase Receptor Type 2 (NTRK2) Gene Associated with Treatment Response to Mood Stabilizers in Patients with Bipolar I Disorder. *J Mol Neurosci.* 2013 Jun; 50(2): p. 305-10.

[89] Zhang JP, Lencz T, Geisler S, Derosse P, Bromet EJ, et al. Genetic variation in BDNF is associated with antipsychotic treatment resistance in patients with schizophrenia. *Schizophr Res.* 2013 May; 146(1-3): p. 285-8.

[90] Baum AE, Akula N, Cabanero M, Cardona I, Corona W, et al. A genome-wide association study implicates diacylglycerol kinase eta (DGKH) and several other genes in the etiology of bipolar disorder. *Mol Psychiatry.* 2008 Feb; 13(2): p. 197-207.

[91] Yosifova A, Mushiroda T, Kubo M, Takahashi A, Kamatani Y, et al. Genome-wide association study on bipolar disorder in the Bulgarian population. *Genes Brain Behav.* 2011 Oct;10(7):789-97. 2011 Oct; 10(7): p. 789-97.

[92] Ollila HM, Soronen P, Silander K, Palo OM, Kieseppä T, et al. Findings from bipolar disorder genome-wide association studies replicate in a Finnish bipolar family-cohort. *Mol Psychiatry*. 2009 Apr; 14(4): p. 351-3.

[93] Squassina A, Manchia M, Congiu D, Severino G, Chillotti C, et al. The diacylglycerol kinase eta gene and bipolar disorder: a replication study in a Sardinian sample. *Mol Psychiatry*. 2009 Apr; 14(4): p. 350-1.

[94] Zeng Z, Wang T, Li T, Li Y, Chen P, et al. Common SNPs and haplotypes in DGKH are associated with bipolar disorder and schizophrenia in the Chinese Han population. *Mol Psychiatry*. 2011 May; 16(5): p. 473-5.

[95] Weber H, Kittel-Schneider S, Gessner A, Domschke K, Neuner M, et al. Cross-disorder analysis of bipolar risk genes: further evidence of DGKH as a risk gene for bipolar disorder, but also unipolar depression and adult ADHD. *Neuropsychopharmacology*. 2011 Sep; 36(10): p. 2076-85.

[96] Tesli M, Kähler AK, Andreassen BK, Werge T, Mors O, et al. No association between DGKH and bipolar disorder in a Scandinavian case-control sample. *Psychiatr Genet.* 2009 Oct; 19(5): p. 269-72.

[97] Takata A, Kawasaki H, Iwayama Y, Yamada K, Gotoh L, et al. Nominal association between a polymorphism in DGKH and bipolar disorder detected in a meta-analysis of East Asian case–control samples. *Psychiatry Clin Neurosci*. 2011 Apr; 65(3): p. 280-5.

[98] Moya PR, Murphy DL, McMahon FJ, Wendland JR. Increased gene expression of diacylglycerol kinase η in bipolar disorder. *Int J Neuropsychopharmacol*. 2010 Sep; 13(8): p. 1127-8.

[99] Britsch S. The neuregulin-I/ErbB signaling system in development and disease. *Adv Anat Embryol Cell Biol*. 2007; 190: p. 1-65.

[100] Green EK, Raybould R, Macgregor S, Gordon-Smith K, Heron J. Operation of the Schizophrenia Susceptibility Gene, Neuregulin 1, Across Traditional Diagnostic Boundaries to Increase Risk for Bipolar Disorder. *Arch Gen Psychiatry*. 2005 Jun; 62(6): p. 642-8.

[101] Prata DP, Breen G, Osborne S, Munro J, St Clair D, et al. An association study of the neuregulin 1 gene, bipolar affective disorder and psychosis. *Psychiatr Genet*. 2009 Jun; 19(3): p. 113-6.

[102] Goes FS, Willour VL, Zandi PP, Belmonte PL, MacKinnon DF. Family-Based Association Study of Neuregulin 1 With Psychotic Bipolar Disorder. *Am J Med Genet B Neuropsychiatr Genet.* 2009 Jul; 150B(5): p. 693-702.

[103] Cassidy F, Roche S, Claffey E, McKeon P. First family-based test for association of neuregulin with bipolar affective disorder. *Mol Psychiatry*. 2006 Aug; 11(8): p. 706-7.

[104] Crisafulli C, Chiesa A, Han C, Lee SJ, Balzarro B, et al. Case-control association study of 36 single-nucleotide polymorphisms within 10 candidate genes for major depression and bipolar disorder. *Psychiatry Res*. 2013 Aug; 209(1): p. 121-3.

[105] McDonald ML, MacMullen C, Liu DJ, Leal SM, Davis RL. Genetic association of cyclicAMPsignaling genes with bipolar disorder. *Transl Psychiatry*. 2012 Oct; 2: p. e169.

[106] Patel SD, Le-Niculescu H, Koller DL, Green SD, Lahiri DK, et al. Coming to grips with complex disorders: genetic Risk Prediction in Bipolar Disorder Using Panels of Genes Identified Through Convergent Functiopnal Genomics. *Am J Med Genet B Neuropsychiatr Genet.* 2010 Jun; 153B(4): p. 850-77.

[107] Blair IP, Chetcuti AF, Badenhop RF, Scimone A, Moses MJ, et al. Positional cloning, association analysis and expression studies provide convergent evidence that the cadherin gene FAT contains a bipolar disorder susceptibility allele. Mol Psychiatry. 2006 Apr; 11(4): p. 372-83.

[108] Abou Jamra R, Becker T, Georgi A, Feulner T, Schumacher J, et al. Genetic variation of the FAT gene at 4q35 is associated with bipolar affective disorder. *Mol Psychiatry.* 2008 Mar; 13(3): p. 277-84.

[109] Belmonte Mahon P, Pirooznia M, Goes FS, Seifuddin F, Steele J, et al. Genome-Wide Association analysis of age at onset and psychotic symptoms in bipolar disorder. *Am J Med Genet B Neuropsychiatr Genet.* 2011 Apr; 156B(3): p. 370-8. Erratum in: *Am J Med Genet B Neuropsychiatr Genet.* 2011 Sep;156B(6):p.749-50.

[110] Morris LG, Kaufman AM, Gong Y, Ramaswami D, Walsh LA, et al. Recurrent somatic mutation of FAT1 in multiple human cancers leads to aberrant Wnt activation. *Nat Genet.* 2013 Mar; 45(3): p. 253-61.

[111] Logan CY, Nusse R. The Wnt signaling pathway in development and disease. *Annu Rev Cell Dev Biol.* 2004; 20: p. 781-810.

[112] Terracciano A, Esko T, Sutin AR, de Moor MH, Meirelles O, et al. Meta-analysis of genome-wide association studies identifies common variants in CTNNA2 associated with excitement-seeking. *Transl Psychiatry.* 2011 Oct; 1: p. e49.

[113] Chu TT, Liu Y. An integrated genomic analysis of gene-function correlation on schizophrenia susceptibility genes. *J Hum Genet.* 2010 May; 55(5): p. 285-92.

[114] Lesch KP, Timmesfeld N, Renner TJ, Halperin R, Röser C. Molecular genetics of adult ADHD: converging evidence from genome-wide association and extended pedigree linkage studies. *J Neural Transm.* 2008 Nov; 115(11): p. 1573-85.

[115] Soronen P, Ollila HM, Antila M, Silander K, Palo OM, et al. Replication of GWAS of bipolar disorder: association of SNPs near CDH7 with bipolar disorder and visual processing. *Mol Psychiatry.* 2010 Jan; 15(1): p. 4-6.

[116] Cichon S, Mühleisen TW, Degenhardt FA, Mattheisen M, Miró X, et al. Genome-wide association study identifies genetic variation in neurocan as a susceptibility factor for bipolar disorder. *Am J Hum Genet.* 2011 Mar; 88(3): p. 372-81.

[117] Miró X, Meier S, Dreisow ML, Frank J, Strohmaier J, et al. Studies in humans and mice implicate neurocan in the etiology of mania. *Am J Psychiatry.* 2012 Sep; 169(9): p. 982-90.

[118] Soria V, Martínez-Amorós E, Escaramís G, Valero J, Pérez-Egea R, et al. Differential association of circadian genes with mood disorders: CRY1 and NPAS2 are associated with unipolar major depression and CLOCK and VIP with bipolar disorder. *Neuropsychopharmacology.* 2010 May; 35(6): p. 1279-89.

[119] Shi J, Wittke-Thompson JK, Badner JA, Hattori E, Potash JB, et al. Clock genes may influence bipolar disorder susceptibility and dysfunctional circadian rhythm. *Am J Med Genet B Neuropsychiatr Genet.* 2008 Oct; 147B(7): p. 1047-55.

[120] Mansour HA, Talkowski ME, Wood J, Chowdari KV, McClain L, et al. Association study of 21 circadian genes with bipolar I disorder, schizoaffective disorder, and schizophrenia. *Bipolar Disord.* 2009 Nov; 11(7): p. 701-10.

[121] Kishi T, Kitajima T, Ikeda M, Yamanouchi Y, Kinoshita Y, et al. Association study of clock gene (CLOCK) and schizophrenia and mood disorders in the Japanese population. *Eur Arch Psychiatry Clin Neurosci.* 2009 Aug; 259(5): p. 293-7.

[122] Roybal K, Theobold D, Graham A, DiNieri JA, Russo SJ, et al. Mania-like behavior induced by disruption of CLOCK. *Proc Natl Acad Sci U S A*. 2007 Apr; 104(15): p. 6406-11.
[123] Mukherjee S, Coque L, Cao JL, Kumar J, Chakravarty S, et al. Knockdown of Clock in the ventral tegmental area through RNA interference results in a mixed state of mania and depression-like behavior. *Biol Psychiatry*. 2010 Sep; 68(6): p. 503-11.
[124] van Enkhuizen J, Minassian A, Young JW. Further evidence for ClockΔ19 mice as a model for bipolar disorder mania using cross-species tests of exploration and sensorimotor gating. *Behav Brain Res*. 2013 Jul; 249: p. 44-54.
[125] Jirtle RL, Skinner MK. Environmental epigenomics and disease susceptibility. *Nat Rev Genet.* 2007 Apr; 8(4): p. 253-62.
[126] Vincent JB, Masellis M, Lawrence J, Choi V, Gurling HM. Genetic association analysis of serotonin system genes in bipolar affective disorder. *Am J Psychiatry*. 1999 Jan; 156(1): p. 136-8.
[127] Kishi T, Okochi T, Tsunoka T, Okumura T, Kitajima T, et al. Serotonin 1A receptor gene, schizophrenia and bipolar disorder: an association study and meta-analysis. *Psychiatry Res.* 2011 Jan; 185(1-2): p. 20-6.
[128] Carrard A, Salzmann A, Malafosse A, Karege F. Increased DNA methylation status of the serotonin receptor HTR1A gene promoter in schizophrenia and bipolar disorder. *J Affect Disord.* 2011 Aug; 132(3): p. 450-3.
[129] Sugawara H, Iwamoto K, Bundo M, Ueda J, Miyauchi T, et al. Hypermethylation of serotonin transporter gene in bipolar disorder detected by epigenome analysis of discordant monozygotic twins. *Transl Psychiatry*. 2011 Jul; 1: p. e24.
[130] Asai T, Bundo M, Sugawara H, Sunaga F, Ueda J, et al. Effect of mood stabilizers on DNA methylation in human neuroblastoma cells. *Int J Neuropsychopharmacol*. 2013 Nov; 16(10): p. 2285-94.
[131] Massat I, Kocabas NA, Crisafulli C, Chiesa A, Calati R, et al. COMT and age at onset in mood disorders: a replication and extension study. *Neurosci Lett*. 2011 Jul; 498(3): p. 218-21.
[132] Benedetti F, Dallaspezia S, Colombo C, Lorenzi C, Pirovano A, et al. Association between catechol-O-methyltransferase Val (108/158)Met polymorphism and psychotic features of bipolar disorder. *J Affect Disord*. 2010 Sep; 125(1-3): p. 341-4.
[133] Abdolmaleky HM, Cheng KH, Faraone SV, Wilcox M, Glatt SJ, et al. Hypomethylation of MB-COMT promoter is a major risk factor for schizophrenia and bipolar disorder. *Hum Mol Genet.* 2006 Nov; 15(21): p. 3132-45.
[134] Nohesara S, Ghadirivasfi M, Mostafavi S, Eskandari MR, Ahmadkhaniha H, et al. DNA hypomethylation of MB-COMT promoter in the DNA derived from saliva in schizophrenia and bipolar disorder. *J Psychiatr Res*. 2011 Nov; 45(11): p. 1432-8.
[135] Dempster EL, Pidsley R, Schalkwyk LC, Owens S, Georgiades A, et al. Disease-associated epigenetic changes in monozygotic twins discordant for schizophrenia and bipolar disorder. *Hum Mol Genet*. 2011 Dec; 20(24): p. 4786-96.
[136] Severinsen JE, Als TD, Binderup H, Kruse TA, Wang AG, et al. Association Analyses Suggest GPR24 as a Shared Susceptibility Gene for Bipolar Affective Disorder and Schizophrenia. *Am J Med Genet B Neuropsychiatr Genet*. 2006 Jul; 141B(5): p. 524-33.
[137] Chaki S, Funakoshi T, Hirota-Okuno S, Nishiguchi M, Shimazaki T, et al. Anxiolytic- and antidepressant-like profile of ATC0065 and ATC0175: Nonpeptidic and orally

active melanin-concentrating hormone receptor 1 antagonists. *J Pharmacol Exp Ther.* 2005 May; 313(2): p. 831-9.

[138] Borowsky B, Durkin MM, Ogozalek K, Marzabadi MR, DeLeon J, et al. Antidepressant, anxiolytic and anorectic effects of a melanin-concentrating hormone-1 receptor antagonist. *Nat Med.* 2002 Aug; 8(8): p. 825-30.

[139] Kaminsky Z, Tochigi M, Jia P, Pal M, Mill J, et al. A multi-tissue analysis identifies HLA complex group 9 gene methylation differences in bipolar disorder. Mol Psychiatry. 2012 Jul; 17(7): p. 728-40.

[140] Veldic M, Guidotti A, Maloku E, Davis JM, Costa E. In psychosis, cortical interneurons overexpress DNA-methyltransferase 1. *Proc Natl Acad Sci U S A.* 2005 Feb; 102(6): p. 2152-7.

[141] Guidotti A, Ruzicka W, Grayson DR, Veldic M, Pinna G, et al. S-adenosyl methionine and DNA methyltransferase-1 mRNA overexpression in psychosis. *Neuroreport.* 2007 Jan; 18(1): p. 57-60.

[142] Higuchi F, Uchida S, Yamagata H, Otsuki K, Hobara T, et al. State-dependent changes in the expression of DNA methyltransferases in mood disorder patients. *J Psychiatr Res.* 2011 Oct; 45(10): p. 1295-300.

NEWS

In: Biochemistry and Molecular Biology
Editor: Mary Miu Yee Waye

ISBN: 978-1-63482-312-8

Chapter 9

CONFERENCE REPORT OF 2012 INTER-UNIVERSITY POST-GRADAUTE SYMPOSIUM HELD IN HONG KONG

Mary M. Y. Waye

The Division of Life Sciences, The Hong Kong University of Science and Technology hosted the third Inter-University Post-Graduate Symposium 2012. This event was jointly organized with sister departments at The University of Hong Kong (HKU) and The Chinese University of Hong Kong (CUHK) with the aim of promoting inter-institutional and inter-disciplinary communications among postgraduate students and faculty members. The keynote speakers were Prof. Min HAN HHMI, Univ. of Colorado at Boulder, USA and Prof. Xi HE Children's Hospital Boston/Harvard Medical School, USA. Three awards for Oral Presentation (HK$1,000) and ten awards Poster Presentation (HK$500) were given this year.

Source of Photos: Annie Au, HKUST. Details of the event (including photos) can be found in this web site:http://biopgsymp.ust.hk/cgi-bin/index.php

SCIENTIFIC PROGRAMME

9:15 Min HAN HHMI, Univ. of Colorado at Boulder, USA You are what you eat: 10 year study of the roles of lipid variants, miRNA and metabolism in animal development and stress response to starvation and pathogen

10:15 Kit Ying Crystal Leung CUHK Sox9 is important for determining neuron-glial fate choice in mouse cerebellar development

10:50 Hoi Leong Xavier Wong HKU. The Regulation of FGF Signaling by MT1-MMP in Calvarial Osteogenesis

11:10 Qing LU HKUST Mechanistic study of myosin X for filopodial formation

11:30 Minghui GAO CUHK The Mechanisms of Apoptosis in Human Erythrocytes

11:50 Jie CHEN HKU RNF169 ANTAGONISES THE UBIQUITIN-DEPENDENT SIGNALING CASCADE AT SITES OF DNA DAMAGE

12:10 Yang SHEN HKUST Melanocortin-4 receptor-mediated regulation of hippocampal synaptic plasticity

14:00 Xi HE Children's Hospital Boston/Harvard Medical School, USA Understanding Wnt signaling and the Wnt Morphogen in development and disease

15:00 Ho Chun WONG CUHK Assembly of preactivation complex for urease maturation in Helicobacter pylori: crystal structure of UreF-UreH protein complex.

16:20 Ka Yiu KONG HKU Yeast TRAMP complex facilitates pre-mRNA splicing

16:40 Xing GUO HKUST miR-203: a new tumor suppressor in Rhabdomyosarcoma

17:00 Award Ceremony

ABSTRACTS PRESENTED

Mr. Kar Lun Gallant Chan	The study of the removal methods and the origin of nitrite in edible bird's nest
Mr. Hoi Leong Xavier Wong	The Regulation of FGF Signaling by MT1-MMP in Calvarial Osteogenesis
Miss Weng Hong Sou	Characterization of the Interaction and Phosphorylation of Serine/Arginine-rich Splicing Factor 3 by SR Protein Kinase 2
Miss Minghui Gao	The Mechanisms of Apoptosis in Human Erythrocytes
Miss Si Xie	UHRF1 Double Tudor Domain and the Adjacent PHD Finger Act Together to Recognize K9me3-Containing Histone H3 Tail
Miss Wing Heng Conny Yu	Structural Characterization of Human Ribosomal Stalk Proteins
Mr. Shang Yuan	Guanylate kinase domains of the MAGUK family scaffold proteins as specific phospho-protein-binding modules
Mr. Jie Chen	RNF169 Antagonises the Ubiquitin-dependent signaling cascade at sites of DNA damage
Mr. Yan-Hui Fan	Genome-Wide Association Study of Adolescent Idiopathic Scoliosis in Southern Chinese
Mr. Hing Yuen Yeung	Photodynamic therapy in cancer treatment
Miss Yuen Ting Wong	Maize ribosome-inactivating protein interacts with P2 via Lys158-Lys161 and the activity of RIPs is related to their interaction with ribosome
Mr. Derek Cheung	PinX1 is involved in telomerase recruitment and regulates telomerase function by mediating its localization
Miss Alice Kar Lai Yang	The Combination of Loop-mediated Isothermal Amplification (LAMP), Gold Nanoparticles and qNano technology for MRSA DNA detection
Mr. JIn Xiwan	Effects of trophic stimulation on the maturation of the neuromuscular junction
Miss Ka Yee Au	Targeted killing of HIV/SIV infected cells by modified maize ribosome-inactivating protein with viral protease recognition sites
Miss Fei YE	INAD PDZ2 plays a regulatory role in the Drosophila visual signaling.
Miss Pengfei Liu	Characterizing the Association of NME7with the 帚-Tubulin Ring Complex
Mr. Guangyu Guo	Relative and accurate measurement of protein abundance using 15N stable isotope labeling in Arabidopsis (SILIA)
Mr. Xing Guo	miR-203: a new tumor suppressor in Rhabdomyosarcoma
Mr. Yang Shen	Melanocortin-4 receptor-mediated regulation of hippocampal synaptic plasticity
Miss Lei Yang	Expression of RGS20 enhances cell aggregation and is associated with the metastatic potential of cancer cells
Ms. Yuan yuan Yu	Developing a new approach to degenerative disc disease therapy: DNA aptamers to inhibit aggrecanases

Mr. Ka Leong Ao	Dissecting the mechanism of binding between polyphosphate and polyphosphate kinase 2
Mr. Lin Zhu	Functional phosphoproteomics analysis of novel ethylene signaling component ERF110
Mr. Chao Wang	Structure of the ZU5-ZU5-UPA-DD tandem of ankyrin-B reveals interaction surfaces necessary for ankyrin function
Mr. Ka Ho Raymond Yip	Identification and Characterization of Long-range SOX9 Enhancers in Limb Development
Miss Wan Ning Vanessa Chow	FE65 interacts with huntingtin and modulates mutant huntingtin degradation via the ubiquitin/proteasome system
Miss Yan Wang	EpiRegNet: constructing epigenetic regulatory network from high throughput gene expression data for humans
Mr. Xu Feng	A fast and accurate SNP detection algorithm for next-generation-sequencing data
Miss Qing Lu	Mechanistic study of myosin X for filopodial formation
Miss Tsz Ying Lee	Protein Stability in Ionic Liquids
Mr. Ka Kui Tong	Co-Smad independent BMP canonical Smad signaling is required for the specification process of the anterior rhombic lip during cerebellum development
Mr. Ricky Wong	Expression and functional study of foxp4 in the central nervous system of zebrafish
Mr. Ho Chun Wong	Assembly of preactivation complex for urease maturation in Helicobacter pylori: crystal structure of UreF-UreH protein complex.
Mr. Hao Zuo	Regulation of Fhit by Gq family proteins
Miss Yuchen Liu	The Roles of Irx3 and Irx5 Genes in Mammalian Inner Ear Development
Mr. Wang Chihao	Substrate Recognition Mechanism of Atypical Protein Kinase Cs Revealed by the Structure of PKC in Complex with a Substrate Peptide from Par-3
Miss Jin Zhou	Loss of MT1-MMP influences CNS development
Miss Yee Ling Lam	Involvement of Proheparanase and Peri-synaptic Heparan Sulfate Proteoglycan in Synaptic plasticity to regulate AMPA-Type Glutamate Receptor Trafficking
Ms. Yun Shen	Extracellular matrix and stem cell niche in the regenerating planarian
Mr. Ka Yiu Kong	Yeast TRAMP complex facilitates pre-mRNA splicing
Dr. Kit Hang Chu	Sox10 is Essential for Neural Crest Invasion in Cochlea-Vestibular Ganglion
Mr. Wing Tai Tung	Aligned chitosan nano-fibers for nerve regeneration
Miss Ho Yan Leung	Involvement of Neuregulin 1 Type III in the Ex Vivo Generation of Fate Committed Schwann Cells from Bone Marrow Stromal Cells
Miss Ho Yan Leung	Involvement of Neuregulin 1 Type III in the Ex Vivo Generation of Fate Committed Schwann cells from Bone Marrow Stromal Cells
Miss Man Ning Hui	Investigating the role of Sox9 in human neural stem cells
Mr. Lap Kei Kenneth Wu	Heparanase Activity Facilitates Syndecan-1 Shedding in Airway Inflammation
Mr. Kwok Lam Fung	Activity assays of Chondroitinase ABCI immobilized on chitosan beads

Mr. Wesley Tucker	DNA Aptamers as inhibitors of E3 Ubiquitin ligase WWPI
Ms. Hong Huan Hor	Suppressor of fused (Su(fu)) Plays Domain Specific Function in Orchestrating The Maintenance and Differentiation of Hindbrain Neural Progenitor Pools
Mr. Ian Chong	The Use of Small Molecules to Direct the Differentiation of Embryonic Stem Cells into Insulin-secreting β-like cells
Mr. S H L	Spatial Temporal Automatic Microscope
Ms. Chen Feng	Role of IHH signaling in the progression of digit joint formation
Miss Tai Wing Yin Evelyn	Fate commitment of bone marrow-derived Schwann cells
Ms. Cui Guo	Dynamics of phytoplankton community structure in the South China Sea in response to the East Asian aerosol input
Mr. Weixin Wang	Next generation sequencing has lower sequence coverage and poorer SNP-detection capability in the regulatory regions
Mr. Hao Zhang	The Genetic and Phenotypic Effects of Coliphage on E. coli
Ms. Huiyan Gan	Kinesin-1 is necessary for cytokinesis in chondrocytes
Miss Wing Yan Hung	The role of TAX 1 Binding Protein 2 in hepatocellular carcinoma.
Mr. Hari Krishna	A Novel Neural Response Algorithm for Protein Function Prediction
Miss Wing Yan Yu	Progenitor Cells in the Posterior Limbus: An Attractive Source for Stem Cell Therapy in Corneal Endothelial Diseases and Glaucoma
Mr. Siu Wong Chung	The role of Cdc48p in the regulation of DNA replication in Budding Yeast
Mr. Xibin Lu	Quantitative characterization of mouse embryonic stem cell self-renewal, differentiation and dedifferentiation
Mr. Hu Jianjiang	An Improved Method For Cell Traction Force Measurement
Mr. Hon Wing Luk	Investigating the involvement of PI3K/PDK-1/Akt pathways and Notch signal in SARS-CoV Membrane-induced apoptosis
Mr. Yue Jia	Systematic Search for EB1-binding Proteins

ACKNOWLEDGMENTS

The 2012 Inter-University Post-Gradaute Symposium was generously sponsored by Chinetek Scientific (China) Ltd. and Eppendorf China Ltd.

In: Biochemistry and Molecular Biology
Editor: Mary Miu Yee Waye
ISBN: 978-1-63482-312-8

Chapter 10

NEWS IN BRIEF. I INTERVIEW WITH FIVE MEMBERS OF THE SCHOOL OF BIOMEDICAL SCIENCES, THE CHINESE UNIVERSITY OF HONG KONG

Interview of Prof. Yang Chao Chen

Prof. Yang Chao Chen, PhD, School of Biomedical Science, Chinese University of Hong Kong

Professor YangChao Chen is currently an assistant professor at the School of Biomedical Science in the Chinese University of Hong Kong. His research interests are in the field of functional genomics and molecular biology of tumour and he focuses on gene regulation in gastrointestinal cancer, the development of novel therapeutics for cancer and RNAi-based experimental therapy for HIV and influenza. Currently, he is investigating the function of a particular gene on cancer, its role in contributing to the development of cancer and whether it can be used as a therapy to treat cancer.

The goal of his research is to explore an innovative way to fight influenza and cancer. At the time being, he is focusing on investigation in liver and pancreatic cancer. His researches were being published in several medical journals, for example, Gastroenterology, Hepatology, Cancer Research, etc.

Prof. Chen got his postdoctoral training in University of Washington, Seattle, USA. He got his Ph.D in Sun Yat-sen University in Guangzhou.

In 2009, Prof. Chen acquired a US$100,000 Grand Challenges Explorations grant from the Bill & Melinda Gates Foundation due to his research, being the second CUHK scientist to receive the grant after Prof. Samuel Sai-ming Sun. His research, titled 'Generation of influenza-resistant chicken by triple combination lentiviral vector-mediated genetic modification' proposes in developing a lentiviral vector that targets the entry and replication of influenza viruses in domestic chickens. It is being considered to be falling outside the current scientific paradigms and might be leading to significant advances in global health.

Prof Chen is decidedly optimistic to the future development in science because of the good foundation in this field. Yet he hopes that the Hong Kong government would provide more resources into the field of cancer biology. Being an assistant Professor at the Chinese University, Prof. Chen found that Hong Kong students are diligent and hard-working, and the graduate students in his team perform very well in the research. At last, he encouraged students to pursue into the science field if they are interested as he believes that interest is one of the most important considerations for studying in university.

By Vanessa Ho and Angel Chu; Edited by Kenny Chan

Bibliography: A press release from the Chinese University of Hong Kong, dated 5th of May, 2009

Interview of Dr. Isabel, S. S. Hwang

Dr. Isabel, S. S. Hwang, *B.Sc. (1st Class Hons, London), M.Sc. (Chemical Research, London), DIC. (London), Ph.D. (HK),* School of Biomedical Science, Chinese University of Hong Kong

Dr. Isabel Hwang currently works as a lecturer and is teaching Pharmacy, Chinese Medicine and Nursing courses in the School of Biomedical Science in the Chinese University of Hong Kong. Her major duty as a lecturer is to provide didactic lectures and other learning activities like practical workshops and tutorials to undergraduates.

At the time being, she is focusing on developing animated courseware packages for her students. She has made a number of automarking online exercises and study-related mobile applications for her students to enhance their understanding on the subject and to practice what they have learnt. Her various courseware packages have been recognized by a commendation award given by CLEAR EXPO in 2010.

Dr. Hwang was also awarded a Commendation Prize for her presentation on "Online delivery of pre-captured lectures via Echo360 for student learning" delivered at the Teaching and Learning Innovation EXPO 2012 on 13 Dec 2012. The event was co-organized by the Centre for Learning Enhancement and Research (CLEAR) and the Information Technology Services Centre (ITSC).

Dr. Hwang had obtained two degrees of Biochemistry in London, one of them as the first class of Honors. Comparing her past experience with the life of the Hong Kong students nowadays, she believes that the greatest difference between her college life and theirs is the kinds of exposure she had had during her times studying in the universities. Sharing some of her experience, she mentioned that even though she is studying organic chemistry, she had also been required to explore other fields like arts. Moreover, she said that learning in London allowed her to access more resources and to learn to improve herself during the time of studying. Speaking of her students, she described the majority of her students as hard-working but at the same time not adventurous enough that they are 'limiting' themselves. She found that her students depend heavily on their teachers' handouts and notes, yet they do not have the self-innovation to read other related materials. She encouraged Hong Kong graduates to be more talkative, not in the sense of talking to each other when the Lecturer is teaching, but to be brave to ask questions. At the same time, she hoped that her students will be more initiative and willing to approach subjects no matter they are included in syllabus or not.

By Angel Chu and Vanessa Ho; Edited by Kenny Chan

Bibliography: http://mmlab.itsc.cuhk.edu.hk/eNewsASP/app/article-details.aspx/24B9AB1D7DE312D739C58B53829D8730/ (dated 2011/6/22)

Interview of Prof. Sydney SB YU

Professor Sidney Yu, B.A., Ph.D., *School of Biomedical Science, Chinese University of Hong Kong*

Being a science researcher for years, Prof. Yu has returned to Hong Kong three years ago and serves as a research assistant professor in the Chinese University of Hong Kong. In addition to offering education in cell biology and specializes cell organelle, he conducts research in transport of vesicles from the nucleus through ERGIC to the target destination.

Inspired by his mother, Prof. Yu has been interested in cell biology since he was a high school student. Later, he continued his study overseas in America. Having the experience of studying abroad, he realized that the greatest obstacle faced by most overseas students is to adapt to an English learning environment. The bright side of studying overseas is that students are more proactive and have stronger interactions during lessons. Yet, he sees no big difference between local and overseas universities.

Research is not an easy task. The most exciting moment during research is when hard work paid off. Sometimes the experimental results have shown remarkable advances although most of the time the results may not be as encouraging. In fact, researchers have to deal with difficult times when funding resources for the pursuit of science is limited. On the whole, he thinks that keeping on his scientific work is worthwhile.

To students, Prof. Yu believes that one has to be passionate about the science field, as well as being careful when carrying out experiments. Student should bestow much time on exploring the science world but not too exam-oriented. He suggests reading more reference books may help one to be more prepared for the capricious science.

For those interested in cell biology, Prof. Yu stressed that students should have enthusiasm for learning science. His final remark for the privileged generation nowadays is to treasure the chance of pursuing our dreams.

By Joyce Wong and Amy Fong
Edited by Kenny Chan

Interview of Prof. Hui Zhao

Professor Zhao Hui, *B.Sc., M.Sc, Ph.D., School of Biomedical Sciences, the Chinese University of Hong Kong*

Professor Zhao Hui completed his studies in the Mainland China. He worked for five years in Germany after getting his master degree and then went abroad to join the National Institute of Health in the United States. He is currently a research assistant professor at the Chinese University of Hong Kong. His work focuses on early embryonic development while his research interests also involve the signalling networks in the neural crest formation and the mechanisms of protein trafficking in regulating FGF signalling.

Research on early embryonic development focuses on the molecular mechanisms of the embryo after fertilization, including how the embryo becomes a body and how the molecules work together. The findings are fundamental scientific knowledge. Everyone will benefit as we can then understand more about our physical development. The current project is a study of genes. This type of research has utmost significance to society as it can help explain human development and could eventually lead to the prevention of birth defects. Moreover, the results are applicable to many other areas of study such as cancer research as the cells involved are similar.

Professor Zhao has spent the past three years at CUHK. During which he has found that Hong Kong students are very hard-working, studious and eager learners. He encourages students to venture into the science field and become scientists. Yet, it requires strength and bravery. Research consists of trial and error and repeated experiments, which could all lead to frustrations. Thus, scientists need to have enough courage to face and overcome these obstacles, which could then lead to great scientific breakthroughs.

He deeply appreciates the new school of biomedical sciences at CUHK. This has allowed more collaboration and facilitated communication between different departments.

As for the future development of science in Hong Kong, he believes that the Government should inject more resources into the research field. This would further improve the quality of research.

By Saidal Chan and Eloisa de Vera
Edited by Kenny Chan

Interview of Prof. G Xiao Hua Jian

Professor Jiang Xiao Hua, *School of Biomedical Science, Chinese University of Hong Kong*

After years of cancer and stem cell research in the United States, Prof. Jiang returned to Hong Kong for further exploration on cancer and stem cell biology and she is currently an assistant professor in the Chinese University of Hong Kong. She has great enthusiasm in discovering how the molecular mechanisms controlling normal stem cell function are hijacked by the cancer cells to generate tumor. Through these studies, Prof. Jiang believes her team will make a substantial and significant contribution to the development of stem cell-targeted cancer therapies, thus advancing the filed towards its ultimate goal of improving survival and quality of life for cancer patients.

Recently, Prof. Jiang has found that some stem cell genes and pathways are aberrantly expressed by cancer cells, and play critical role in the development of cancer. Though the long-term investigation is tough and persistent, she can still acquire a sense of satisfaction in the process. Inevitably, there may be some false hypothesis that leads the research to the wrong path; however, obtaining desirable results is always an encouragement to scientists to carry on the research.

Compared with her career in USA, she pointed out that the experience here in Hong Kong is different. Research resources in USA are rather rich and the Government provides sufficient support to the scientists. Moreover, there are no boundaries between different science departments, in contrast, less communication but more independent science sectors are found in Hong Kong.

From Prof. Jiang's point of view, studying science requires a sensitive sense to science issues and even every detail in our daily lives. With prudent observation, even tiny and minor things can be a clue for investigation. Moreover, students should always bear an innovative mindset to explore science.

By Carrie Lam and Mona Tam
Edited by Kenny Chan

In: Biochemistry and Molecular Biology
Editor: Mary Miu Yee Waye
ISBN: 978-1-63482-312-8

Chapter 11

My Role in the Development of the Department of Biochemistry, the Chinese University of Hong Kong

Kai Keung Mark
Dept. of Biology, The Chinese University of Hong Kong and Dept.
Dept. of Science, Hong Kong Institute of Education

Introduction

Historically, Biochemistry emerged from research areas between chemistry and biology, and this was also true for the formation of the Department of Biochemistry in The Chinese University of Hong Kong. The starting nucleus of Biochemistry Board came from three research groups within the three colleges of the university around the period of 1969-1971. First, Prof. Ma LIN (1) of Chemistry Department, United College (2) which was located in Caine Road, Hong Kong side at that time. Prof. Ma worked on proteins and hormones. Second, Prof. Kung Hing YUNG (3) and Prof. Yuen Cheung KONG of Biology Department, Chung Chi College (4), in Shatin. Prof. Kong worked on animal physiology and Prof. Yung worked on plant physiology. Third, myself (Prof. Kai Keung MARK) of the Department of Biology, New Asia College (5) which was located in Farm road, Kowloon at that time. My research interest was on Molecular Biology with an emphasis on DNA and bacteriophages.

When we were still separated in different campuses and with different laboratory facilities during such time (1969-71), The Chinese University started a centralization process, so my Biochemistry course become an inter-collegiate course of the Biology Board between Chung Chi College and New Asia College with lectures in a central location in Star House at Tsim Sha Tsui with Prof. Yung and Prof. Kong also teaching some subjects. Then the Biology and Chemistry Boards jointly proposed to set up a new working group, the Biochemistry working group involving Prof. Ma, Prof. Kong, Prof. Yung and myself as the panel chairman of this working group. Prof. Ma and myself worked out a detail plan to promote biochemistry as an Inter-Board program between Biology Board and Chemistry

Board. Prof. Ma took a successful sabbatical leave to Prof. Choh Hao Li's laboratory (6) in San Francisco Medical Center to work on growth hormones. On his return, he was designated as the founding chair of the new Biochemistry Board in the first floor of the south wing of the new Science Center, and during those days (1971-1974), I used the research laboratory of Rm193, and Prof. Kong was in Rm194, and Prof. Yung was in Rm195, while Prof. Ma used the conference room converted research laboratory opposing these three laboratories. The research facilities came from the existing Biology Board such as Beckman ultra-centrifuge and spectrophotometers. The basic teaching program, Biochemistry, was taught by these four basic members, as well as other newer members including Prof. Cheuk Yu LEE and Prof. Hin Wing YEUNG. As the Biochemistry Board expanded with a bigger budget, more laboratory space, and more full-time teaching staff, it also recruited undergraduate and graduate students. Prof. Ma did a wonderful administrative job in handling the transition including the laboratory transfer of Prof. Yung and myself who originally belonged to the Biology Board. I led the move to G96 on the ground floor, and retained a close and harmonious relationship with Biochemistry Board afterwards.

As Prof. Ma became Vice-Chancellor of the University, and Prof. Lee became the head of the Biochemistry Board; there were many ups and downs in conflicts between the Biology and Biochemistry Board, due to competition for resources and laboratory space. I was lucky to avoid these confrontations, and retained as a chelating element between these two competing Boards.

Up to now, I repeatedly use the term Biology and Biochemistry Board, instead of the more common term "department", this is because The Chinese University of Hong Kong is composed of colleges, and experienced a painful merging process, so before the merging process is completed, the term "department" belonged to the colleges, while the term "board" belonged to the University, so I have to use the confusing term "Board'. From now on the merging has completed, and the term "department" no longer has any college implications, so I will switch back to the use of Biochemistry Department.

My second phase of involvement in the development of Biochemistry Department is related to the modern advances in gene cloning and biotechnology. In 1984, Prof. Ma, still Vice-Chancellor of The Chinese University of Hong Kong, helped me to obtain a Croucher Foundation Grant to spend my sabbatical leave at the Institute of Biotechnology, Imperial College of Science and Technology, London, UK, to learn the cloning technology under the supervision of Prof. Brian Hartley.

On returning to The Chinese University of Hong Kong, I teamed up with Prof. Walter Ho, my good friend, to establish a joint Biology-Biochemistry Departmental Biotechnology laboratory in Choh Ming Li Basic Medical Sciences Building. This facility enabled staff and graduate students from both departments to pull their research resources and their biotechnological experience together, and operate continuously after its establishment until more recently when the department of Biochemistry was split between the Faculty of Science and the Faculty of Medicine in 2009.

Although I retired as a teaching staff by 1995, but I still join the luncheon gatherings organized by the staff members of the previous department of biochemistry, and the Biochemistry staff also treat me as one of their gang members, and invites me for their functions.

This long association for over thirty years has many basic reasons behind it. Firstly, I am a biochemist with a Ph. D. degree in Chemistry, so I can comfortably get involved in their scientific discussions. Secondly, many of the teaching staff members are my former students, and thirdly most members also have interest in my tricks in making money from the stock market (7), and the bets between us to invite all staff members to join a banquet if The Hang Seng Index reached a certain level provided us with some exciting entertainment. All these factors keep me within this tightly knit group of scientists.

REFERENCES

[1] "Professor Ma Lin- Biography" Press release of Shaw College, The Chinese University of Hong Kong. Available from: http://www.cuhk.edu.hk/ipro/pressrelease/ma_lin-e.htm
[2] Official home page of United College. Available from: http://www2.cuhk.edu.hk/uc/eng/
[3] "Professor Prof. Kung Hing YUNG" Press release of The Chinese University of Hong Kong. Available from: http://www.cuhk.edu.hk/ipro/pressrelease/040514Yung.htm
[4] Official Home page of Chung Chi College. Available from:http://www.ccc.cuhk.edu.hk/index.php?lang=en
[5] History of New Asia College, Available from: http://www.na.cuhk.edu.hk/index.php?option=com_contentandview=articleandid=1andItemid=2andlang=en
[6] Biographical memoirs of Li Choh-Hao by David Cole. Available from http://www.nap.edu/readingroom/books/biomems/cli.html
[7] Mark's Column. Available from: http://www.asiawind.com/pub/hksr/mark/mark1.shtml

In: Biochemistry and Molecular Biology
Editor: Mary Miu Yee Waye
ISBN: 978-1-63482-312-8

Chapter 12

Meeting Report: 6th European Summer School in Proteomics Basics- Brief Report

R. Dineshram
The Swire Institute of Marine Science and School of Biological Sciences,
The University of Hong Kong

As a PhD student working on the oyster larvae looking at their impact due to climate change, I required more knowledge demanding molecular techniques like advance proteomics to deliver fact and consequences of climate change in my study. It was an unforgettable experience and knowledgeable trip for me in Italy, having the opportunity to attend the European summer school course to learn more about basic to advance proteomics.

The workshop was held in a monastery at Bressanone, Italy from August 19-25, 2012 in a place surrounded by a beautiful landscape full of vineyards and apple orchards. The summer course had participants from different countries, and I was one among three students got travel grant funded by FEBS, UK under "FEBS Advanced Lecture Course- High Performance Proteomics" with the support of Hong Kong Society of Biochemistry and Molecular Biology (http://www.bch.cuhk.edu.hk/hksbmb/).

As a marine biologist and PhD student, my project involved in studying the effects of chronic exposure levels to CO_2 concentrations of current atmospheric levels, pH 8.2 and predicted future levels to pH 7.9 and pH 7.6 thereby predicting the potential impact of climate change on oyster aquaculture in South China. The workshop provided me a unique opportunity to get more insights on the proteomics technologies starting from basics to advanced technologies developed in the field of life sciences. The workshop materials were highly useful providing fresh impetus in applying proteomics in marine biology field of research.

The workshop was run with many lectures from invited guest speakers introducing some new concepts including how to design an experiment, prepare samples and analyze the data with computational tools. It also had sessions talking about little advancement in post translational modification strategies and integrating "omics" data which is a way forward understanding biology in the field of science. Apart from the lectures and workshop, the participants were also made to involve in other extracurricular activities like mountain biking and hiking.

Since after this workshop it got me more confidence in my research that paved way for helping me to publish a paper related to my research in Marine biology (Dineshram et al, Marine Biology 2013) using proteomics tools. Overall the summer school was a joyous moment making me meet the best people in my field of research with a meaningful experience in my research career.

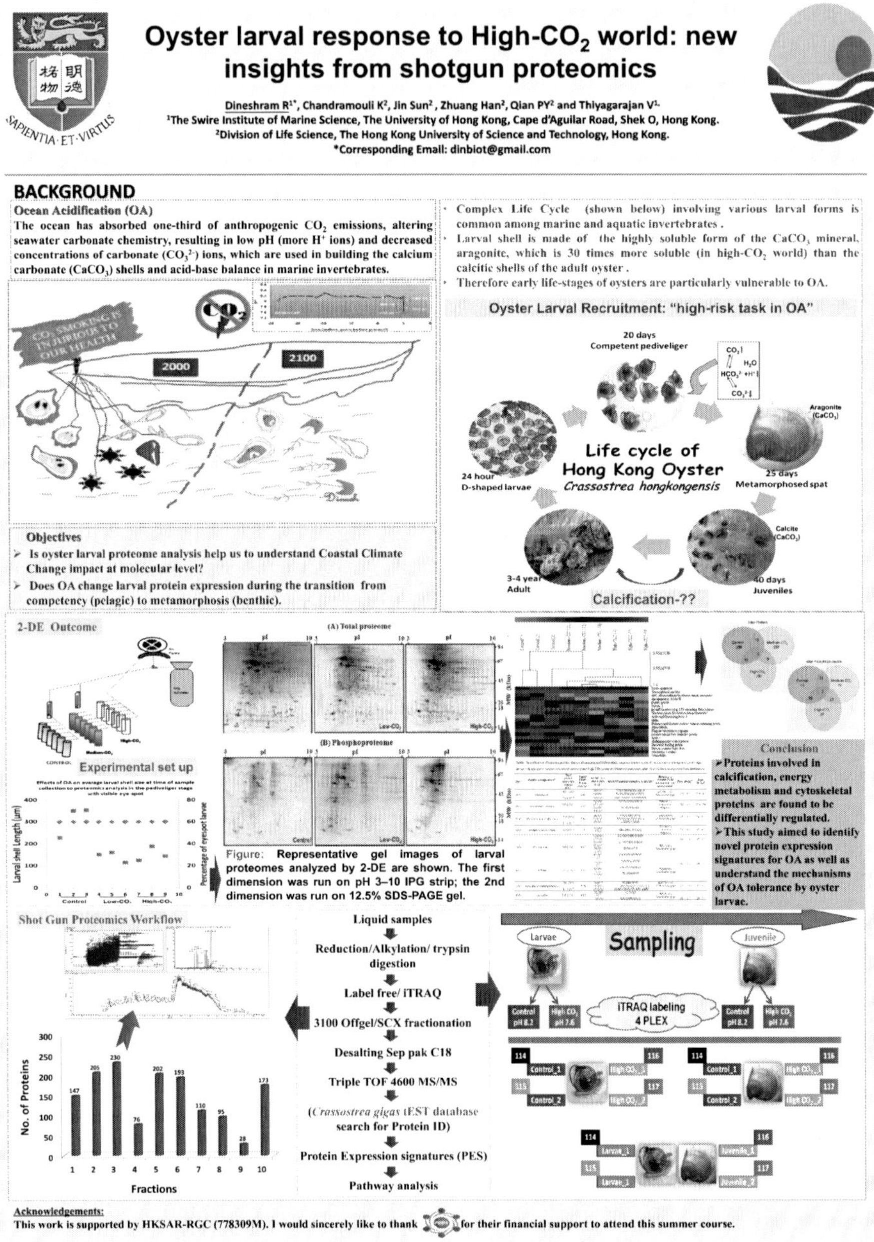

My poster presented at 6th European Summer School in Proteomics Basics.

Mr. Dineshram with his poster during the poster session

Group photo with the summer school participants

Hiking organized by the Summer school

Meeting with John Cottrell, Mascot protein search engine developer

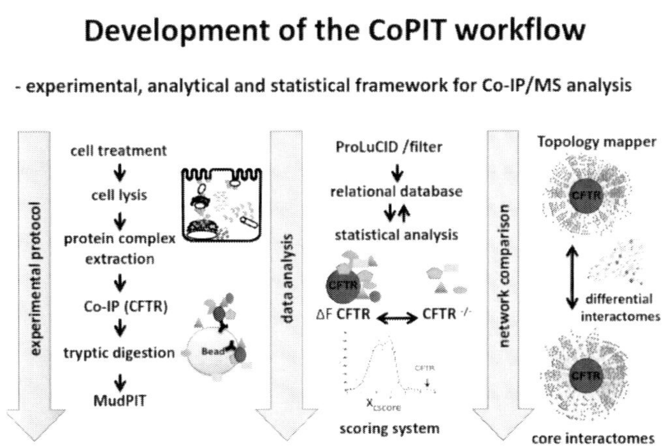

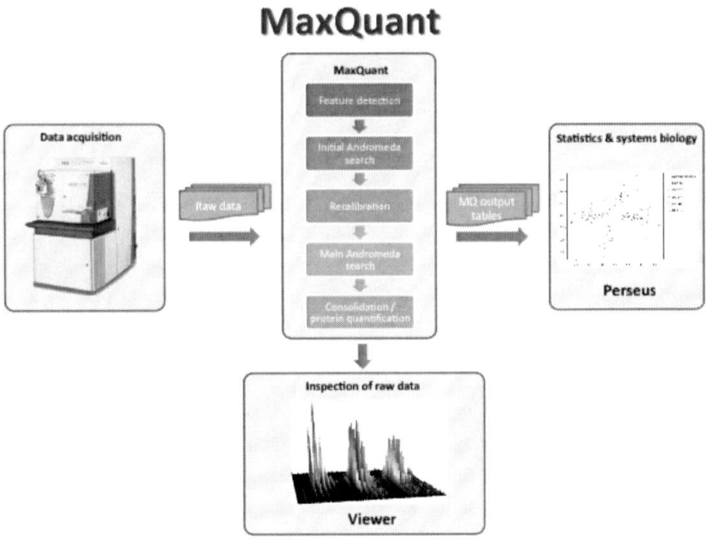

Interactive lectures and workshops to gain more experience on different proteomics tools.

REFERENCES

[1] Ko KWG, Chan VBS, Dineshram R, Choi DKS, Li AJ, et al. (2013) Larval and Post-Larval Stages of Pacific Oyster (Crassostrea gigas) Are Resistant to Elevated CO_2. *PLoS ONE* 8(5): e64147. doi:10.1371/journal.pone.0064147.

[2] Dineshram R, Lane A, Wang K, Xiao S, Yu Z, et al.. (2012) Analysis of Pacific oyster larval proteome and its response to high-CO_2. *Mar Pollut Bull.* In Press.

[3] Dineshram R, Thiyagarajan V, Lane A, Ziniu Y, Xiao S, et al.. (2013) Elevated CO_2 alters larval proteome and its phosphorylation status in the commercial oyster, Crassostrea hongkongensis. *Mar Biol* (In Press).

In: Biochemistry and Molecular Biology
Editor: Mary Miu Yee Waye

ISBN: 978-1-63482-312-8

Chapter 13

News in Brief. II
Interview with Four Members of the Chinese University of Hong Kong

*An Interview with Professor Jacky Ngo: Determination and Passion by Vanessa Yeung**

(Good Hope School)

Background

Professor Jacky Ngo is an Assistant Professor of School of Life Sciences of The Chinese University of Hong Kong. His research interests are the structure-function studies of proteins that involve the regulation of pre-mRNA splicing, the roles of slicing kinases in various cancers and viral infections and the molecular basis of interactions between and their substrates.

* vanessaysw@gmail.com

Research Is by Searching Repeatedly

Recently, Prof. Ngo is conducting a research on the drug screening project against HIV1 and a collaborative research on the drug discovery, for instance, Spinocerebellar Atrophy. "Research is by searching repeatedly", said Prof. Ngo emphatically with sparkling eyes. Errors like reagent contamination, environmental errors and personal errors are commonly found in researches, which make every research unforgettable. Moreover, overcoming challenges plays an important role in doing research. For example, it might be quite difficult to predict the result such as structures related to X-ray. Another challenge is that not every research project can be completed. "50% of a successful process of protein crystallography depends on luck", said Prof. Ngo with a grudging smile. Despite the outcome is always uncertain, Prof. Ngo has never thought of giving up. I could feel his enthusiasm in doing research and no obstacle or discouragement can reduce his passion in his research work.

We Need More Support

With his expertise in conducting researches and his professional knowledge in scientific research works, Prof. Ngo compared the research funding in US and in Hong Kong. Undoubtedly, the former gains more support from the government and companies like biomedical equipment firms. As the demand for chemicals and equipment assisting researches in the US is relatively higher than that in Hong Kong, discounts are usually provided to university laboratories which contribute lowering the cost of doing research. However, this is not the case in Hong Kong as subsidies provided by HK government is insufficient to support the research and on the other hand, high shipping fees and commissions paid to distributers doubled the burden to research works of universities in Hong Kong.

Self-Interest and Inspiration

"My determination in becoming a professor is inspired by my teacher and my interest in researching acted as catalyst throughout my journey of pursing my dream", said Prof. Ngo with adamant eyes. Prof. Ngo had once served as a teaching assistant in the US where he provided real classroom teaching in university. This experience gained him interest in educating youngsters and doing research. The encouragement from his teacher is another major force in achieving his dream as he would like to inspire his students like how he was inspired by his teacher when he was a student. With his passion in passing knowledge to students and enthusiasm in striving for excellence, he reaches where he is today.

Be Determined and Confident

When asked of requirements of being a professor, Prof. Ngo answered with no doubt, "the ability of facing failure". "There may be consistent failures in doing researches, we should face them positively and bear in mind that research is by searching repeatedly. We should always be ready for obstacles in doing research", said Prof. Ngo. Moreover, he remarked that one should be able to learn from mistakes. "It is acceptable to make mistakes but most importantly, one should learn from them so as to avoid repeating the same error."

STUDENTS FROM HONG KONG, MAINLAND AND FOREIGN COUNTRIES

Prof. Ngo evaluated students from Hong Kong, mainland and foreign countries in terms of diligence, ability to cope with difficulties and creativity. Hong Kong students are the most flexible in dealing with difficulties than that of students from mainland and foreign countries. However, they are relatively less hard-working than the mainland students and less creative than students from foreign countries. On the other hand, mainland students are the most diligent and students from foreign countries are the most innovative. It implies that Hong Kong students are as competitive as students from mainland or foreign countries but more determination and passion in striving improvement are needed.

*Interview with Professor Wong Chi Hang Eric by Jory Fong**

BACKGROUND

Prof. Wong Chi Hang Eric, is recently working as an assistant professor at the Prince of Wales Hospital. He mainly focuses on the researches on tumors and cancers in the Department of Clinical Oncology. The Department of Clinical Oncology aims to function as "one of the leading Comprehensive Cancer Centers in South East Asia and Southern China" by applying innovative diagnostic techniques, development of patients support services and clinical laboratory research, as well as teaching undergraduate and post-graduate students". Prof. Wong graduated in 2004 and obtained his Ph.D. in the Department of Biochemistry at the Chinese University of Hong Kong. He spent 2 years working on infectious diseases after his graduation. Then, he went to Singapore for 4 years to do research on cell biology under the supervision of Prof. Li Hoi Yeung. He went back to Hong Kong in 2011 to work on cancer cell related research, such as drug resistant mechanisms in the hospital.

EVERYDAY WORK

Prof. Wong's everyday work can be divided into 2 types, namely translational research and drug testing. First, translational research is a linkage between basic science principle and clinical studies. He collaborates with the doctors in the hospital to study on some special

* <jory_fong0107@yahoo.com.hk>

cases of cancer patients, for example, the tumour in the patient's body divides uncontrollably after taking treatments or 2 drugs needed to be taken together in order to magnify its efficacy. The cancer cells and how treatments work on the cells are different in every patient. Therefore, doctors and scientists need to do follow up on each patient case by case. Prof. Wong is doing some scientific research to back up doctors' clinical studies and explain the phenomenon based on their diagnosis.

Second, scientists in the cancer centre are also doing some drug testing unit that collaborates with the drug companies. Drug companies will send new drugs to the hospital and scientists will carry out a series of testing to find out whether the drug can successfully minimizes the division of cancer cells. If there is a positive result, it will be developed to a new type of medicine and benefits the patients. Prof. Wong reminds us that the cancer cells may have different mutations and exists in different forms in different countries. Therefore, different treatments should be done for cancer patients from Asia and Europe. The drug testing unit can be classified into 3 stages. In the first stage, scientists study on the molecular level. They observe the activity on the tumour under a microscope after applying the drugs on it. When it shows a positive result which the tumour stops growing uncontrollably, the testing will proceed to stage 2. In the second stage, scientists first inject certain types of cancer cell into the body of an animal, usually mice. After the cancer cells grow into a tumour, drug will then be applied. Scientist study on how well the drug can control the division of cancer cells. If it also shows a positive result, the testing will proceed to the final stage. In the final stage, scientists apply the drug on human's body. They will select cancer patients with suitable condition to carry out the test. Doctors look closely to the clinical response of the patients. If the drug can pass all the stages in the drug testing unit, a new medicine can be released to the market to help more cancer patients.

DIFFICULTIES

When we asked Prof. Wong about the difficulties that he faced in his research, he replied, "Everyday is a challenge, basically we face difficulties in every moment." Doing research is not an easy task, it is a long journey which requires much perseverance and determination of scientists. This is because scientists need to discover a new area for research which is not discovered by others yet in the past. In the beginning, scientists have to brainstorm some basic ideas for study. After that, preliminary tests need to be done to find out whether these ideas or hypothesis is workable and can be further developed. At last, more detailed research and experiments can be carried out to develop an explanation for some phenomenon, new drugs and new treatments etc. Prof. Wong emphasizes that we should not give up easily whenever we face obstacles because difficulties are something that we can learn from and we can accumulate experience in every trails. What drives us to success is a million times of trails and experience. Also, his team faces technical problems, such as a patient refuses to take the risk to try a newly developed medicine. Moreover, the major problem that scientists face in research is lacking support from the government. The research funding is not enough to compensate the huge expenditure on research and development. Every field should be allocated with a similar amount of resources and funds, so each field does not get much. R&D is not popular in Hong Kong. Thus, scientists in Hong Kong just focus on small projects with limited resources. The career path of studying science is relatively narrow. As a whole, the environment of research is not keen in Hong Kong. Therefore, scientists face many problems both technically and financially every day.

DIFFERENCES BETWEEN STUDENTS IN THE PAST AND NOW

Prof. Wong introduces to us that the Prince Wales Hospital cancer centre admits biomedical or biomedical engineering students to become research students or assistants. He observed that nowadays more and more girls study science and are interested in doing research while boys tend to study engineering or finance. Furthermore, the quality of students also changed a lot. Students in the past were more hardworking and treasure every opportunities of studying. Students nowadays are less hardworking, but they are very innovative and creative. They have many ideas when giving them the same question. However, they fail to follow up after brainstorming the ideas and their problem solving skills is quite low. Some of them easily give up when facing difficulties and do not find other ways to solve the problem. Also, their ability of managing feelings is poor. Some of them are not willing to accept others' opinion or criticism from those who are experienced. They are quite self-centered and not humble enough when learning.

FEELINGS AFTER VISITING THE LAB AND DOING THE INTERVIEW

I was amazed by the scale of the laboratory in the cancer centre. There are the most advanced equipments that I have never seen before. These equipments aim to give a more accurate result on research. After the interview, I realized that doing research is not as easy as I expected. Scientists need to pay attention to every single detail in the progress of research because a tiny mistake will affect the whole result. Also, interest and perseverance is very important in doing research

Interview with Professor Jerome Hui Ho Lam by Lauren Lam

POSITIVITY AND ENTHUSIASM

As soon as my partner and I entered the office where the word '3:30 interview' was written on the white board, Professor Hui of the School of Life Sciences, The Chinese University of Hong Kong, immediately greeted us with a big smile on his face. 'What are you two planning to study at University?' As we were settling down for the interview, Professor Hui suddenly sprang a question on us. He is quite concerned about the impact of the current education system on teenagers.

'Though the new DSE (Diploma of Secondary Education) system may bring a lot more choices to students, it may also adversely affect them in that many actually do not exactly know what they want to do in the future.' He observed, after listening to both of our indistinct answers. He expressed that many Hong Kong students nowadays lack clear goals and are not able to focus on their true interests, as they are given too many choices. Furthermore, more and more teenagers in the face of increased choices and competitions will choose to study in certain fields only because it ensures impressive job prospects, rather than their own interests.

'It is certainly a great concern because people cannot really do well when they study something that they are not interested in.' he said with a frown. Being too pragmatic is becoming one of the characteristics of Hong Kong students, compared with students from other countries. 'They are just too pragmatic that sometimes the attitude diminishes the true meaning of learning.' He continued. Using his unique and extensive experiences as supportive evidence, he believes that interest should be the most important factor considered when they are choosing their studies. After all, the main reasons for him to become a professor are his enthusiasm in doing researches and passion in education.

Being a Professor

Professor Hui dreamt of being a scientist since early childhood. As he grew up, he excelled in Biology which further confirmed his road in studying science. 'I believe that the meaning of life is to use one's own talent to contribute to the society and world.' Considering his impressive academic results in Biology and affable personality, he decided to become a professor.

The road to Professor is indeed long and tortuous. It requires a lot of intensive studying and extensive work experiences. Also, academic achievements during studies must stand out from others so to showcase one's ability and build up reputation. Other than academic ability, the character and personality are also important in paving the road to Professor. One has to be passionate in conducting research but also enjoys teaching the younger generations in order to become a good professor. And last but not least, perseverance is the key. If one does not have perseverance, even though he was extremely talented, he would still be unable to endure the long hours and dedication needed to become a professor.

'It's all about persistence, passion and enthusiasm!' He concluded.

Research

Interested in his life as Professor, we continued our interview asking him about his research. Though he smiled shyly claiming that he had just become a professor for half a year (as of July, 2013) and was inexperienced, he still explained patiently about his work to us.

The main field of Professor Hui is evolution biology. He mainly investigates into the foundational mechanisms of non-coding RNA and their applications. His research is mainly categorized into 4 parts: how the functions of non-coding RNA affect animal development and evolution; how they contribute to the embryonic development; how they affect the hormonal response; and the interpretation of genomic sequences.

When asked about his most unforgettable experience in research, he suddenly became silent. Thinking that he had not heard us clearly, we repeated the question. This was when he smiled saying, 'Don't worry, I heard the question. I am just wondering which one as all experiences are indeed unforgettable for me.' After a while, he recalled two experiments. One was about his research in Oxford. About eight years ago, a young man doing research on the gene organization of invertebrates, never expected there would be surprising results: He discovered that the DNA sequence of a certain insect was actually quite similar to that of the human. It was more unforgettable that he had failed many times in this experiment before he succeeded. When he was recounting on the experience, he sighed that it was a difficult time as he was very poor and could not afford the expensive research. 'From this experience, I learnt the importance of perseverance. It is always fine to fail as long as you learn from the failure and be mindful not to repeat the error again in future. This is why human can improve.'

POSITIVITY

It is easy to see that Professor Hui is a very positive person. Talking about contributions, he told us that all research has its contribution to society. 'All research could help to enlarge the boundary enclosing human's knowledge; the only difference is the extent of the enlargement.' Indeed, even basic research may help to build up the foundation of a field, which would be applicable to future research and contributes to future development. After all, research is not just about attaining knowledge about life. How to apply it in future development is also crucial in any scientific development. 'The potential of any discovery may be massive, so all research contributes to society,' He added.

After this, the discussion moved to another topic --- research funding, which is a good example of Professor Hui's positive attitude. Research funding in Hong Kong has relatively been lower than that of other countries. According to a comparative study conducted by Hong Kong Institution of Science (HKIS) on the development of science and technology in Hong Kong and other regional economies[1], the research funding of Hong Kong by far lags behind other countries like Taiwan and Singapore. However, Professor Hui is still hopeful about the situation. 'Though research funding in Hong Kong is comparatively lower than other countries, Hong Kong have many talented students. There is hope for improvement in Hong Kong if we do something to improve Hong Kong's science development.' He said with his eyebrows raised with a bright expression.

Since research is a long journey which requires much time and effort. If you do not have an interest on it, it is very hard to continue with it. I appreciate all the scientists who are working hard on the road of research with limited support from the government.

[1] Hong Kong Institute of Science --- Research and Development in Hong Kong: The Way Forward. Retrieved from http://www.science.org.hk/index.php?class=index&action=reports_1 on 4th July, 2013

Report on the Research of Professor Wong Chun Kwok

*by Sharon Chung**

BACKGROUND

Professor Wong Chun Kwok obtained his PhD degree in Immunology from the Department of Biochemistry, CUHK in 1993, postdoctoral appointments at the Centre for Biochemical and Biophysical Science and Medicine of the Brigham and Women's Hospital, Harvard Medical School, USA (1994-1996), and the Pulmonary and Critical Care Unit at the Massachusetts General Hospital, Harvard Medical School (1996-1997). Afterwards, he was appointed postdoctoral research in the Department from August 1997, Scientific Officer in July 2000, Associate Professor in 2001 and Professor since 1 August 2009. He has also been appointed as Assistant Director of Institute of Chinese Medicine, CUHK since 2009 and Joint State Key Laboratory of Phytochemistry and Plant Resources in West China since 2010. Professor Wong has been the President of Hong Kong Society for Immunology and Hong Kong Society of Flow Cytometry and current Executive Member of the Hong Kong Society for Immunology and Institute of Biology, Hong Kong Branch.

Professor Wong is also involved in teaching medical students for Immunology, Clinical Biochemistry and General Education and even some evening classes for part-time students. He has mentioned that Immunology is the core subject which all medical students in CUHK have to study in Year 2 and Biochemistry is a study on the chemical reactions inside human body. However, general education is not related to science as it is more about common sense. Students are assigned with projects of free topic and have presentation during class. Instead of teaching, Professor Wong spends most of his time in research include Immunology, Chinese

* <sharon_chung709@yahoo.com.hk>

Medicine etc. He also serves as Honorary Scientific Officer (Medical) in the diagnostic clinical chemistry laboratory at Prince of Wales Hospital.

DETAILS OF THE RESEARCH

Professor Wong spent most of his time on research work and he has worked for 30 years since 18, his first research was about Chinese Medicine. For the research he has done, the research can be mainly categorized into two themes: Immunology and Chinese Medicine.

Immunology

His immunology research includes the elucidation of cytokine and chemokine network and intracellular signal transductions of immune effector cells including eosinophils, lymphocytes, mast cells, basophils, dendritic cells, in inflammatory diseases, with recent emphasizing in pattern recognition receptors mediated innate immunity and damaging inflammation, the immunopathological mechanisms of allergic diseases, autoimmune diseases, infectious diseases and diabetes mellitus, with interdepartmental collaboration and the immunomodulatory, anti-tumor and anti-allergic activities of traditional Chinese medicine.

From the interview, Professor Wong has provided a more detailed explanation on his research. First, allergic diseases are the common chronic diseases which require a long period of time to recover and difficult to cure, asthma and allergic rhinitis are the common examples for allergic diseases. The foreign substances that cause allergies are called allergens and mostly they can be found from the surrounding environment. Dust mites, pollen, animal hair, insect strings are some common allergens and dust mites would be the most common allergens among all of them. Dust mites are too small to see without a microscope and usually suspended in air. It has 4 pairs of limbs and eats skin cells shed by people. Dust mite will lead to an overreaction of immune system and cause allergy occurs. The immune system misinterprets it as a disease agent and begins producing a type of antibody against it, called immunoglobulin E (IgE). The IgE produced during the primary response binds to basophils in the bloodstream and to a similar type of cell called mast cells in the tissues. When the person again encounters the allergen, these basophils and mast cells that have bound to IgE release histamine, prostaglandins and leukotrienes, which cause inflammation of the surrounding tissues, resulting in allergic symptoms.

Furthermore, he has also done a series of research on autoimmune diseases, infectious diseases and diabetes mellitus. He gave well-explained description on every disease and provided specific examples for a better understanding. For autoimmune diseases, he explained that it means an inappropriate immune response of the body against substances and tissues normally present in the body such as kidney, Systemic Lupus Erythematosus (SLE) is one of the examples of autoimmune disease and usually occurs in female. After that, he started to give an explanation on diabetes mellitus. Diabetes mellitus occurs when the pancreas does not produce enough insulin, or when the body cells cannot respond to the insulin produced. This results in a highly variable blood glucose level and can cause severe damage to the blood vessels of the heart, kidneys, eyes and nerves etc after a long period of time. There are two forms of diabetes mellitus, type 1 diabetes and type 2 diabetes. Type 1 diabetes results from little or no insulin is produced since the insulin-secreting cells are destroyed by the immune system. Type 2 diabetes results from insulin resistance, which body cells are insensitive to insulin. The term "type 1 diabetes" has replaced several former terms, including childhood-onset diabetes, juvenile diabetes, and insulin-dependent diabetes mellitus (IDDM). Likewise, the term "type 2 diabetes" has replaced several former terms, including adult-onset diabetes, obesity-related diabetes, and noninsulin-dependent diabetes mellitus (NIDDM).

Chinese Medicine

Instead of the research on immunology, he has also studied a series of research on Chinese Medicine. He has mentioned that the first research that he has done was about Chinese Medicine, by the research on Chinese Medicine, he believes that Chinese herbs can improve the body immune system such as Panax ginseng, antler, angelica etc. From his research, his team has developed a Chinese Medicine "三冬涼茶" which can help to prevent influenza.

Views towards the Students

He believes that there are differences between the students in the past and nowadays. In the past, mostly 100% of the students are Hong Kong citizens, while there are 70% of students come from mainland China nowadays. Cantonese would be the mother language of all students in the past, however, it has changed to mandarin nowadays since the students are mainly come from China. However, the attitude towards studying has been improved for the mainland students put more effort on studying while those in the past did not. Since the living environment of Hong Kong students and that of mainland students are relatively different, Hong Kong students enjoy a better living environment when compare to the others. Therefore, they may spend more time on leisure activities rather than in studying. However, the living environment of the mainland students are comparatively poor, they choose to spend more time on academic as they do not have the opportunities to have fun with friends and resulted in paying more attention to their studies and get a better results when compare to that of Hong Kong students. Although the mainland students have put more effort than the Hong Kong students, their language skills are not good enough (especially English skills). For being a university student in Hong Kong, English is very important as most of the subjects use English for teaching. Also, part of the subjects may include oral presentation or communication which requires English skills. Since there is less education on English in China, the mainland students may not have much chance to use English. Thus, the English skills are not good enough especially on speaking during examination.

Problems Faced

Professor Wong has mentioned that research is a difficult work which requires lots of time and money. Research funding would be the biggest problem that he has faced as Hong Kong is relatively difficult to get funding from the government. In Hong Kong, there are lots of people working on different research and resulted in a more competitive situation. Moreover, the research funding from Hong Kong government is not sufficient enough for all kinds of research. The Hong Kong government only provides 600-700 thousands Hong Kong dollars per two to three years. This amount is completely not enough for all the research as the money required for a research is relatively large, say as much as around 1 million per year. As the research funding is not enough, they have to use different ways to get more money like approaching some sponsors. National Natural Science Foundation of China is one of the sponsors which provide financial support for science research. Professor Wong has recommended that the scientists should get as much money as possible from different sources, so to provide a sufficient amount of money to continue their own research.

INDEX

C

D

J

K

L

M

N

O

P

Q

W

Y